Applied Engineering
Failure Analysis

APPLIED ENGINEERING FAILURE ANALYSIS

Theory and Practice

Hock-Chye Qua • Ching-Seong Tan • Kok-Cheong Wong
Jee-Hou Ho • Xin Wang • Eng-Hwa Yap
Jong-Boon Ooi • Yee-Shiuan Wong

CRC Press
Taylor & Francis Group
Boca Raton London New York

CRC Press is an imprint of the
Taylor & Francis Group, an **informa** business

CRC Press
Taylor & Francis Group
6000 Broken Sound Parkway NW, Suite 300
Boca Raton, FL 33487-2742

First issued in paperback 2017

© 2015 by Taylor & Francis Group, LLC
CRC Press is an imprint of Taylor & Francis Group, an Informa business

No claim to original U.S. Government works

ISBN-13: 978-1-4822-2218-0 (hbk)
ISBN-13: 978-1-138-74786-9 (pbk)

Library of Congress Cataloging-in-Publication Data

Qua, Hock-Chye.
    Applied engineering failure analysis : theory and practice / authors, Hock-Chye Qua, Ching-Seong Tan, Kok-Cheong Wong, Jee-Hou Ho, Xin Wang, Eng-Hwa Yap, Jong-Boon Ooi, Yee-Shiuan Wong Hock-Chye Qua and # others.
        pages cm
    Summary: "This book fills the gap between the traditional failure analysis theory and the actual conduct of the failure cases. The book demonstrates the main methodologies that have evolved from the 1970s to date. Calculation of engineering cases and estimation of system stress and strength are demonstrated in the chapters. A wide range of actual cases span a wide array of engineering fields, including power systems, metallurgy, mining, structures, and machines"-- Provided by publisher.
    Includes bibliographical references and index.
    ISBN 978-1-4822-2218-0 (hardback)
    1. Failure analysis (Engineering) I. Title.

TA169.5.Q33 2015
620.1'12--dc23
                                          2015004554

Visit the Taylor & Francis Web site at
http://www.taylorandfrancis.com

and the CRC Press Web site at
http://www.crcpress.com

Contents

Foreword

It fills me with pride and pleasure to pen the foreword for this book. I must say I am privileged to be given this chance as I personally know some of the authors: Ir. Qua Hock Chye who was my lecturer during my undergraduate days and is now an International Collaborative Partner of Universiti Tunku Abdul Rahman and Dr. Tan Ching Seong who was my colleague at Multimedia University and Universiti Tunku Abdul Rahman. My heartiest congratulations to all the authors who have achieved this remarkable feat of having the book published. It is a testament of your passion and dedication to the area of failure analysis as the way forward towards better investigative research and engineering work.

Former British Prime Minister, Winston Churchill, once said, 'Success is based on going from failure to failure without losing eagerness.' While success is the ultimate power that hastens us en route to our goals, it is failure along the way that guides us towards further discovery and leads us to these goals. Science progresses by trying out ideas, disproving earlier conceptions, and gradually getting closer to the truth at the heart of the phenomenon being studied. Thomas Edison tried more than a thousand times before he produced a light bulb that worked. Failure is a wonderful tutor and an even better teacher for the discovery and prevention of future mishaps in all aspects.

The role of an engineer is to respond to a need by building a device, to plan or to create within a certain set of guidelines and specifications. Some designs will fail to perform their given function with a sought-after level of performance. Hence, engineers must struggle to design in such a way as to avoid failure, and most importantly, catastrophic failure that could result in loss of property, damage to the environment, and possible injury or loss of life. Through analysis of engineering disasters and failures, modern engineers learn what to avoid, how to create better designs, and to seek solutions to improve performance with less chance of future failures.

Therefore, I highly commend the authors of this book who have painstakingly researched and compiled a collection of cases on actual failure analysis which serve as valuable lessons for others to study and learn from. Such a generous sharing of knowledge, experiences, and findings augurs well for the future of the engineering industry and I hope that others will take this lead to do something similar for the advancement of knowledge.

May this book achieve its further goal of providing essential reference material and feedback on design processes and thereby contribute to the avoidance of applied engineering failures in the future.

Ir. Prof. Academician Dato' Dr. Chuah Hean Teik
President
Universiti Tunku Abdul Rahman

THE INSTITUTION OF ENGINEERS, MALAYSIA

Bangunan Ingenieur, Lots 60 & 62, Jalan 52/4, P.O. Box 223, Jalan Sultan, 46720 Petaling Jaya
Tel: 603-79684001/4002 Fax: 603-79577678 E-mail: sec@iem.org.my
Homepage: http://www.myiem.org.my

FOREWORD

It is a well known adage that *"one learns more from failures than from successes"*. It is
also true that we learn from experience; as every experience adds more information to
memory that in time reasoning becomes logical. Thus, it is necessary for us to document
our experiences properly and derive lessons from them.

In spite of such efforts however, accidents will unfortunately continue to happen, often
due to reasons beyond our control. Time and time again, structures and machines have
failed without warning, and often with disastrous consequences.

The bottom line is failures are a fact of life. A failure-free system is more of a myth than
a reality. When failures occur, it is important to conduct prompt investigation of these
failures in order to ascertain their causes and take remedial action to prevent their
recurrence.

Applied Engineering Failure Analysis: Theory and Practice seeks to develop awareness in
engineers on the traditional failure analysis theories and the actual conducts of the
failure cases. It provides a systematic analysis that can be implemented in the design
works to prevent future failures.

The authors have meticulously compiled vast information and case studies on the
subject. As a useful record of the experiences of failure analysis of over a 30-year
period, this book is a good archive of information for failure analysts, practicing
engineers, and students of engineering.

As President of The Institution of Engineers, Malaysia, a professional institution
dedicated to promote the science and profession of engineering and to facilitate the
exchange of information and ideas related to engineering, I am indeed proud to be
associated with this publication.

I congratulate the authors for their professionalism and commitment that they brought
to their work. To the authors, well done, and I wish you every success in your future
endeavors.

Ir. Choo Kok Beng FASc
IEM President

Acknowledgments

The authors would like to acknowledge the case study contributors: H.C. Qua, C.S. Tan, C.S. Fok, Profmetal Sdn Bhd, JLQ Enterprise, and Vertitech Sdn Bhd. Also special thanks to Janet Yong for proofreading.

Authors

Hock-Chye Qua obtained a B.E. (Mech) and an M. Eng. Sc. from the Engineering Faculty of the University of Malaya, where he later served as an associate professor until his retirement in 1999. He was one of the pioneer failure investigators in Malaysia and has been active in failure/forensic investigations for more than 40 years on a large variety of cases within Asia, spanning the mechanical, metallurgical, structural steel, and electrical disciplines. He is a professional engineer and a Fellow of the Institution of Engineers Malaysia.

Ching-Seong Tan is an associate professor in the Faculty of Engineering at Multimedia University, Malaysia. He currently serves as Division 8 Head of CIE Malaysia, under TEEAM (The Electrical and Electronics Association of Malaysia). Dr. Tan received the J.W. Fulbright Award in 2012/2013. He is the speaker for the Compact Fluorescent Lamp (CFL) and Fluorescent Lamp (FL) Recycling Program sponsored by a GEF-UNDP grant. He is also the 2014 recipient of the Visiting Scholar Award from the Centre of Interdisciplinary Mathematics and Statistics (CIMS), Colorado State University (CSU).

Kok-Cheong Wong received a B. Eng (Hons) from the University of Malaya in mechanical engineering, an M. Eng. from the Kyushu Institute of Technology Japan, and a Ph.D. from The University of Nottingham Malaysia Campus. He has several years of working experience in a multinational company. He is currently an associate professor at the University of Nottingham Malaysia Campus, delivering undergraduate courses in the introduction to aerospace technology, aerodynamics, finite element analysis, design, and manufacturing. Dr. Wong's research interests are related to fluid flow and heat transfer, particularly in the areas of jet impingement and microfluidic cooling, convection in porous media, and fluid structure interaction.

Jee-Hou Ho is an associate professor at The University of Nottingham Malaysia Campus, delivering undergraduate courses in machine dynamics, robotics, and automation technology. His research interests include vibration energy harvesting, bio-mechatronics, and robotics. Dr. Ho received his B. Eng. (Hons) and M. Eng. in mechanical and production engineering from Nanyang Technological University (NTU), Singapore, and a Ph.D. from The University of Nottingham Malaysia Campus.

Xin Wang is a senior lecturer at the School of Engineering, Monash University Malaysia, where she delivers undergraduate courses in non-destructive testing and inspection. Her research interests include non-destructive evaluation, optical inspection, and finite element analysis.

Eng-Hwa Yap is a senior lecturer at the University College London's School of Energy and Resources, Australia (UCL Australia), where he leads courses on water resources management and energy efficiency and conservation. Dr. Yap's research work is concentrated largely on sustainable system dynamics modeling using renewable energy technologies, especially on the dynamics of such technology integration within the community with a focus on emerging economies. He holds a Ph.D. in marine engineering from the University College London and an undergraduate degree in marine technology from Plymouth University (United Kingdom).

Jong-Boon Ooi received a B. Eng. (Hons) and M. Eng. Sc. in mechanical engineering from the Universiti Tunku Abdul Rahman, Malaysia. He was a lecturer at the Faculty of Engineering and Built Environment, Tunku Abdul Rahman University College from 2011 to 2014. He is currently pursuing a Ph.D. at Monash University Malaysia. His current research interests include fuel additives, combustion of fuels, and internal combustion engines.

Yee-Shiuan Wong is a B. Eng. (Hons) materials and manufacturing engineering graduate from the Universiti Tunku Abdul Rahman, Malaysia. He was trained by Ir. Qua and Ching-Seong Tan. He is now working as an engineer in an MNC specializing in advanced materials.

List of Abbreviations

Chapter 1

a_c	Critical crack size (Equation 1.1)
ASME	American Society of Mechanical Engineers
BF	Brittle fracture
CP	Codes of practice
CUI	Corrosion under insulation
DD	Disruptive discharge
DF	Ductile fracture
EFA	Engineering failure analysis
K_{IC}	Plane strain fracture toughness (Equation 1.1)
LEFM	Linear elastic fracture mechanics
MIC	Microbiologically influenced corrosion
MVC	Microvoid coalescence
NDE	Non-destructive evaluation
PWHT	Post weld heat treatment
QA	Quality assurance
QC	Quality control
σ_c	Fracture stress (Equation 1.1)
SCC	Stress corrosion cracking
SEM	Scanning electron microscope
Y	Function of crack and stress configurations (Equation 1.1)

Chapter 2

a	Crack length (Equation 2.5)
AAS	Atomic absorption spectroscopy
AISI	American Iron and Steel Institute
API	American Petroleum Institute
APS-C	Advanced photo system type-C
BCC	Body centred cubic structure
BSE	Backscattered electrons
CCD	Charge-coupled device

EBSD-IPF	Electron backscattered diffraction—Inverse pole figure
EDS	Energy dispersive X-ray spectrometry
E	Elastic modulus (Equation 2.4)
FA	Failure analysis
FI	Failure investigator
FM	Fracture mechanics
HCF	High-cycle fatigue
ICP-AES	Inductively coupled plasma atomic emission spectroscopy
K	Stress intensity factor (Equation 2.5)
LCF	Low-cycle fatigue
n	Factor of safety (Equation 2.3)
N_f	Cycles to failure for completely reversed loading (Equation 2.4)
N	Cycle (Equation 2.5)
n_i	Number of cycles at a given stress σ_i (Equation 2.7)
N_i	Number of cycles to failure at a given stress σ_i (Equation 2.7)
NDT	Non-destructive testing
OES	Optical emission spectrometer
PMI	Positive material identification
σ_m	Mean stress (Equations 2.2 and 2.3)
σ_a	Variable stress (Equations 2.2 and 2.3)
S_e	Fully corrected fatigue strength at critical location of component (Equation 2.3)
ε_A	Strain amplitude (Equation 2.4)
SE	Secondary electrons
S-N Curve	Wöhler curve (cyclic stress versus logarithmic scale of cycles to failure)
S_y	Yield stress (Equations 2.1 and 2.3)
TEM	Transmission electron microscope
WDS	Wavelength dispersive X-ray spectrometer

Chapter 3

a	Crack size (Equation 3.2)
BS	British Standard
BRT	Bus rail transit
HB	Brinell hardness
HRC	Rockwell hardness
HV	Vickers hardness
LRT	Light rail transit
MRT	Mass rapid transit
I_{xx}	Moments of inertia
K_{Ic}	Fracture toughness

K_{Id} Dynamic fracture toughness
K_I Stress intensity factor
MPI   Magnetic particle inspection
OD   Outer diameter
σ_{Ic} Critical/fracture stress
TWJ   Thermic weld joint
UT   Ultrasonic test

Chapter 4

AGMA   American Gear Manufacturers Association
API   American Petroleum Institute
σ_c Calculated contact stress number (Equation A4.3)
BDF   Below the drill floor
BS   British Standard
b   Face width (Equations A4.1 and A4.5)
b   Face width of tooth (Equation A4.3)
C_p Coefficient depending on elastic properties of materials (Equation A4.3)
C_m Geometry factor (Equation A4.3)
C_f Surface condition factor (Equation A4.3)
C_s Size factor (Equation A4.3)
C_H Hardness ratio factor (Equation A4.4)
C_L Life factor (Equation A4.4)
C_T Temperature factor (Equation A4.4)
C_R Factor of safety (Equation A4.4)
C_o Overload factor (Equation A4.3)
C_v Dynamic factor (Equation A4.3)
d   Pinion pitch diameter (Equation A4.3)
EDS   Energy dispersive X-ray spectrometer
F_t Transmitted tangential load (Equation A4.3)
F_t Transmitted load (Equation A4.5)
HB   Brinell hardness
HRC   Rockwell hardness
HV   Vickers hardness
I   Geometry factor (Equation A4.3)
J   Geometry factor (Equation A4.5)
LS   Long string
MD   Measured depth
P   Diametral pitch (Equation A4.5)
K_c Overload correction factor (Equation A4.5)
K_s Size correction factor (Equation A4.5)

K_m	Load distribution correction (Equation A4.5)
K_v	Dynamic factor (Equation A4.5)
K_L	Life factor (Equation A4.6)
K_R	Factor of safety (Equation A4.6)
K_T	Temperature factor (Equation A4.6)
σ_t	Calculated stress at root (Equation A4.5)
S_{at}	Allowable fatigue bending stress for material (Equation A4.6)
S_{ac}	Allowable contact stress number (Equation A4.4)
S_{ad}	Maximum allowable design stress (Equation A4.6)
S_c	Surface stress factor of material (Equation A4.1)
S_b	Bending stress factor for material (Equation A4.2)
SEM	Scanning electron microscope
SS	Short string
Xc	Speed factor for wear (Equation A4.1)
X_b	Speed factor for strength (Equation A4.2)
Y	Strength factor (Equation A4.2)
Z	Zone factor (Equation A4.1)

Chapter 5

Bph	Blue phase
CB	Circuit breaker
DE	Drive end
DD	Disruptive discharge
DGA	Dissolved gas analysis
EDS	Energy dispersive X-ray spectrometer
ELCB	Earth leakage circuit breaker
EMF	Electromotive force
ER	End ring
FAT	Factory acceptance test
FI	Failure investigator
HV	High voltage
H	Heat loss
I	Current
LV	Low voltage
Megger	Megaohm meter
MVC	Microvoid coalescence
N-DE	Non-drive end
NDT	Non-destructive testing
OLTC	On-load tap-changer
P	Power
PLC	Programmable logic controller

R	Electrical resistance
Rph	Red phase
SEM	Scanning electron microscope
SFRA	Sweep frequency response analysis
TRW	Transition resistor wire
Tx	Transformer
V	Voltage
WPI	Weather protected type I (motor enclosure)
XLPE	Cross-linked polyethylene
Yph	Yellow phase

Chapter 6

ASME	American Society of Mechanical Engineers
A3	Lower-temperature boundary of austenite region at low carbon content (phase diagram)
BS	British Standard
BTF	Boiler tube failures
HV	Vickers hardness
OD	Outer diameter

Chapter 7

ASTM	American Society of Mechanical Engineers
BS	British Standard
CUI	Corrosion under insulation
EDS	Energy dispersive X-ray spectrometer
JIS	Japanese Industrial Standard
PVC	Poly vinyl chloride
RH_{av}	Average relative humidity (Equations 7.4 and 7.5)
RT_{av}	Average temperature
R.C.	Conventional reinforced concrete
RH	Relative humidity
σ'	Effective stress (Equation 7.6)
σ	Total stress (Equation 7.6)
SPT	Standard penetration tests
T	Temperature
TBM	Temporary bench mark
W	Weight loss due to corrosion penetration (Equation 7.1)

t	Time of exposure in years (Equations 7.1 through 7.5)
u	Water pressure (Equation 7.6)
w	Water content (grams of water content in 1 kg of dry air)
W	Corrosion depth per surface (mm) (Equations 7.2 and 7.4)

Chapter 8

β_j	Phase function of a medium at the point of scattering (for crystalline)
CB	Conduction band
C_m	Specific heat of crystals (Equation 8.1)
C_p	Specific heat of particle (Equation 8.2)
D'	Material thickness (Equation 8.3)
DBD	Direct-backscattered-direct
E_c	Critical valence bonding strength (Equation 8.11)
E_i	Incident energy (Equation 8.9)
E_f	Final energy (Equation 8.9)
EM	Electromagnetic
F_c	Critical fluence (Equation 8.8)
FDTD	Finite difference time difference
FI	Failure investigator
I	Laser intensity (Equation 8.2)
ICNRP	International Commission on Non-Ionizing Radiation Protection
ISO	International Organization for Standardization
κ	Thermal diffusivity (Equation 8.1)
k_m	Thermal conductivity of crystals (Equation 8.1)
LID	Laser-induced damage
m	Neutron mass (Equation 8.11)
M	Target atom mass (Equation 8.11)
φ	Angle of particle movement before and after collision (Equation 8.11)
PKA	Primary knock-on atom
P_0	Incoming pulse (Equation 8.3)
Q_{abs}	Absorption efficiency factor (Equation 8.2)
ρ_m	Density of crystals (Equation 8.1)
ρ_p	Density of particle (Equation 8.2)
RID	Radiation-induced damage
R_j	Reflectance (Equation 8.3)
T_c	Critical temperature
t_j	First mean arriving time of photon group to layer j (Equation 8.3)
T_{ij}	Transmittance of a particular photon group (Kernal function)
τ_r	Rectangular pulse length (Equation 8.8)
UV	Ultra-violet

1

Introduction to Failure Analysis

1.1 What Is Failure Analysis?

The term *failure analysis* can be applied to many situations but what we are specifically concerned with here is *engineering failure analysis* (EFA), which may loosely be defined as the objective investigation or analysis into an engineering failure, using the science and analysis of engineering, to reveal the proximate and root causes. The term *forensic analysis*, or *forensic engineering*, describes a similar process but the term *forensic* encompasses matters that relate to the jurisprudence system or the system of laws that exists in a particular area of activity. An engineering failure is said to occur when a component or structure can no longer perform its function safely and reliably, whether through degradation, deformation, wear, or breakage; the conditions would have breached the limit and serviceability states used in design.

EFA is a multidisciplinary activity, as it encompasses the full range of engineering components in use, electronic, electrical, mechanical, structural, metallurgical, and chemical. In complex cases, for example in an airplane crash, teams of investigators with specialized skills are involved. The engineering techniques used are wide ranging and include varying levels of stress analysis, materials quantification, computer simulations, and possibly, physical testing of models. In most cases, the primary aim is to determine whether the applied stresses have exceeded the strength of the failed components. Blind testing is unproductive; an experienced investigator can usually narrow down the mode of failure after visual examination and then direct further efforts towards the suspected origin of failure, so a physical examination is important as soon as possible after the failure occurrence, before evidence is lost. The terms *stress* and *strength* are widely defined in our usage. Stress would include, for example, mechanical, electrical, environmental, and electromagnetic radiation stresses, and strength would correspondingly refer to mechanical and dielectric strength, corrosion and creep resistance, and radiation resistance.

1.2 Importance of Engineering Failure Analysis

Engineering components are designed to operate reliably for a certain life-span, but premature failure can occur as a result of one or more factors, some being more predictable than others. The consequences to both the party responsible and to the victim are manyfold; the responsible party may suffer from large financial losses in addition to civil litigation and criminal indict-ment, while the victim may suffer from financial losses as well as personal injury or death. In some cases the environment also suffers. Three recent cases serve as good examples: firstly, the recall of defective components of a popular brand of vehicle between 2009 and 2011, where a total of 8.5 million vehicles were affected worldwide [1–3]; secondly, the explosion of an oil drilling rig in the Gulf of Mexico in 2010, which resulted in the death of 11 workers, immense environmental pollution due to the oil spill, and heavy financial losses to the company; and thirdly, the meltdown of the Fukushima reactors in Japan in 2011 following a tsunami, which caused radiation pollution over wide areas and also affected the financial position of the owner [4–6]. The root causes in the first case were essentially engineering based, but engineering and human inadequacies were present in the second and third cases. Today, full forensic investigations are required and measures put into place to reduce the chances of similar incidents from occurring in the future.

Investigations of failure incidents in the past have helped greatly in advanc-ing engineering knowledge and making present day engineering compo-nents safer. Examples include early fatigue studies on railway axle failures, and also fracture mechanics studies following the brittle fracture tendencies of the World War II welded Liberty ships. These two failures were due to insufficient knowledge of material behavior at those particular times, but the resulting investigations have enabled fatigue design and fracture mechan-ics design methodologies to become important disciplines that have become codified in many national standards [7–9], which are being routinely used for engineering design. Presently, for highly complex machines like aircraft, partial prototypes are made and tested to failure, and failure modes and locations are analyzed to identify weak areas where engineering analysis has not been able to accurately describe them.

1.3 Root Causes and Types of Stressors and Stresses

The root causes of failures fall into three fundamental categories: deficiencies in design and material selection, inadequacies in manufacturing/installation/ quality control, and inadequacies in operation and maintenance. Recognition of the stressors in operation and their accompanying stresses is fundamental to EFA; the common stressors/stresses are:

1. *Mechanical*: For example, mechanical forces applied through solid contact or hydrostatic pressure applied through fluid contact; both result in static, dynamic, and impact stresses.

2. *Electrical*: Firstly, when a differential voltage or potential (stressor) exists between two points, an electric field (stress) will exist between them. The separating insulating material will be subjected to the electric stress and if the breakdown strength of the material is insufficient to withstand the stress, disruptive discharge (DD) will occur between the two points. Lightning is a visually apparent example involving immense proportions; at the other end of the scale, in the semiconductor and micro-electronics industries, static electronic discharge into the circuits can only be detected under magnification. Secondly, when current (stressor) flows in a conductor as a result of differential voltage, ohmic heat (stress) will be created and the conductor will heat up, possibly to melting point.

3. *Environmental*: Examples are oxygen and moisture (stressors), which give rise to electrochemical and oxidative potential (stresses) that cause corrosion and oxidation; or heat (stressor), which causes elevated temperatures (stresses) that result in creep rupture, softening, and deformation.

4. *Electromagnetic*: These comprise the full electromagnetic spectrum (stressors), ranging from the low-frequency radio waves, to microwaves, infrared, ultra-violet, visible light, X-rays, and the very high-frequency gamma rays. The low-frequency waves have low-radiation energies (stresses) whereas the high-frequency waves have very high energies and are highly penetrative (for example, radioactive material emits high-frequency waves). All, except the very low-energy radio waves, result in radiation stresses that can cause various types of degradation in materials; radiation stresses are also dangerous to living organisms.

1.4 Some Common Failure Modes

Usually encountered failure modes are fracture, fatigue, electrical failure, corrosion, deformation, wear and tear, and radiation damage. Only the first four modes, which are more commonly encountered, are explained below to illustrate the range of problems they may cause.

1.4.1 Fracture

Fracture is the separation of a component into two or more parts, by brittle or ductile mechanisms. These mechanisms have also been called *fast fracture*, by virtue of the fact that they occur relatively quickly when compared with stable cracking processes. In ductile fractures, the applied stresses would have exceeded the yield stress of the material and the component would have attained its maximum capacity; if it had been adequately designed and manufactured, the failure could then be attributed to some unforeseen overstress. In brittle fractures, the applied stresses may be much below the yield stress of the material and the role of excessive stress concentrators and inadequate material properties attain more importance. These two modes of failure have been studied [10] extensively, and little is left unknown about them.

1.4.1.1 Ductile Fracture

Ductile fracture (DF) occurs under plane stress conditions when the von Mises stress of the component exceeds the uniaxial tensile stress of the material. The fractured region usually exhibits visible yielding and plastic deformation and is usually necked at the fracture; the deformation and fracture processes absorb a large amount of energy. If the fracture path is perpendicular to the maximum tensile stress, the fracture surface tends to have a grayish appearance (for steel), but if it is parallel to the maximum shear stress direction (shear failure) the fracture surface tends to have a characteristically smooth, silky appearance (Figure 1.1). Microscopically, the fracture path would have no preferential crystallographic direction and the microstructure would show a large amount of grain deformation (Figure 1.2). The failure process occurs by microvoid coalescence (MVC) and the voids will be seen as dimples under a scanning electron microscope (SEM) (Figure 1.3). Different loading modes (I, II, or III) will give dimples of different shapes [11], a knowledge of which would allow the operative stress system causing failure to be identified.

1.4.1.2 Brittle Fracture

Brittle fracture (BF) occurs under plane strain conditions in a direction perpendicular to the maximum tensile stress, irrespective of whether the loading is Mode I, II, or III. The process is unstable and usually proceeds to completion once initiated; it is a low-energy process. In accordance with linear elastic fracture mechanics (LEFM) methodology, fracture is predicted when the conditions described by the following equation are met [12–14],

$$\sigma_c = \frac{K_{IC}}{Y\left(\pi a_c\right)^{1/2}} \tag{1.1}$$

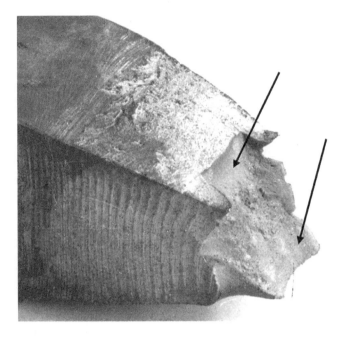

FIGURE 1.1
Shear failure in a fine-grained tensile test bar; the arrows point to fracture surfaces typically described as *smooth and silky*.

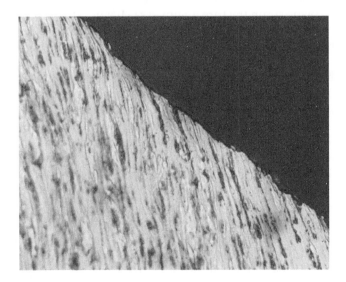

FIGURE 1.2
Microstructure of a shear failure in a fine-grained, ferritic/pearlitic steel; showing a transgranular fracture profile with highly elongated grains. Mag. 500×.

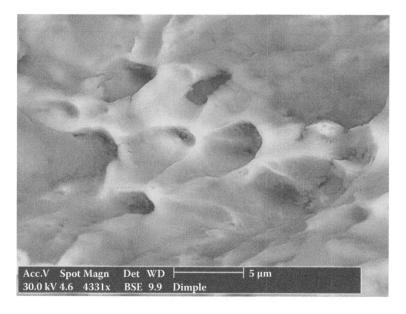

Acc.V   Spot Magn   Det WD ├───────────┤ 5 μm
30.0 kV 4.6   4331x   BSE 9.9   Dimple

FIGURE 1.3
Scanning electron microscope (SEM) fractograph, showing elongated dimples from a ductile shear fracture.

where,

σ_c is the fracture stress (not a constant),
K_{IC} is the plane strain fracture toughness (MPa m$^{-1/2}$), a material property,
Y is a function of the crack and stress configurations,
a_c is the critical crack size (m).

Equation (1.1) shows that the presence of a pre-existing flaw (or stress concentration) is a vital driving parameter; the equation is currently widely used in the design of critical components made of high-strength materials (steels, titanium, and aluminum alloys). The difficulties are in obtaining accurate values of K_{IC}, Y, and a_c for actual components. For the modern, structural ferritic steels where thicknesses in excess of 1 m may be required to attain plane strain conditions, impact energy methods based on ductile to brittle transition behavior are more useful to evaluate whether brittle fracture would be a risk.

Little or no deformation would be visible due to the relatively low stresses involved and the energy consumed would be much lower than that for an equivalent ductile fracture. Macroscopically, small shear lips may be present at the edges and at the end of the fracture surface (Figure 1.4). For steel, the fracture surface would have a bright, shiny, and faceted (grainy) appearance. V-shaped ridges on the fracture surface, commonly called *chevron markings*, point to the crack origin (Figure 1.4). In non-embrittled

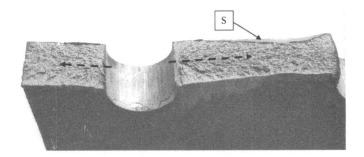

FIGURE 1.4
Brittle fracture of a plate, with chevron marks pointing to locations of crack initiation at the sides of a punched hole. The dashed arrows point to the direction of crack propagation, opposite to the direction of the chevrons; arrow *S* points to a shear lip at the edge of the plate.

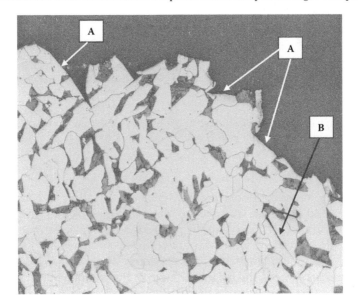

FIGURE 1.5
Microstructure of a brittle fracture profile in a ferritic/pearlitic steel, showing cleavage cracks (A) and a sub-surface cleavage crack (B) at the {100} planes; no grain deformation is visible. Mag. 200×.

ferritic steels, the fracture path travels preferentially along the {100} crystallographic plane by cleavage (Figure 1.5), and little or no deformation would be observed in the material below the fracture profile. SEM fractography will show typical features such as cleavage facets, river patterns, steps, and tongues [11]. In non-embrittled high-strength martensitic steels, failure occurs by quasi-cleavage; if embrittled, intergranular separation would take place along prior austenite grain boundaries and grain boundary facets would be visible under an SEM (Figure 1.6).

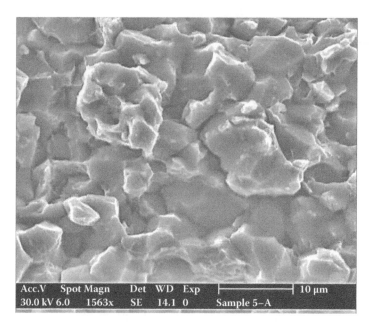

FIGURE 1.6
SEM fractograph of a hydrogen-embrittled, high-strength, martensitic prestressing steel wire, showing predominantly intergranular decohesion along prior austenite grain boundaries.

1.4.2 Fatigue Failure

Fatigue is the most common mode of mechanical failure encountered the world over and occurs in all types of materials; the cracking in the Airbus A380 wing spars (2011) is a highly publicized example. Fatigue failure is caused by stress systems with a variable component, whether of alternating, repeating, or random character. Failure occurs when the magnitude of the applied stresses exceeds the fatigue strength of the component; stress systems may be mechanical in nature or thermally induced. Fatigue can occur over a wide stress spectrum; high stresses cause what is known as low-cycle fatigue (failure in less than 10^3 cycles) and lower stresses cause high-cycle fatigue. In certain materials like steel, there is a stress limit below which fatigue does not occur, an endurance limit. In some materials such as Al and its alloys, no endurance limit is present and the fatigue limit should always be quoted together with the expected cycles to failure. The failure process comprises three stages [15]; crack initiation (Stage I), sub-critical crack propagation (Stage II), and final fracture (Stage III). Stage I occurs on smooth surfaces by slip processes that result in intrusions and extrusions over one or two grain diameters, though this step is often bypassed by the existence of pre-existing flaws. The markings left by this stage are usually obliterated in practical failures and are not given much attention in failure analysis. In Stage II, the crack grows in a stable manner (sub-critical crack propagation) leaving visually

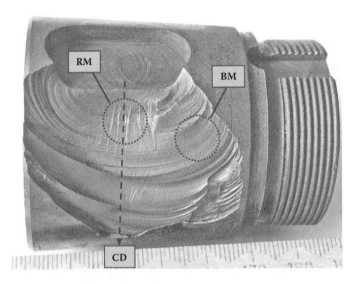

FIGURE 1.7
(See Colour Insert.) Fatigue failure initiating from a sharp edge at the milled keyway in a tapered shaft. Arrow *CD* shows the crack direction, area *BM* the beach marks, and area *RM* the ratchet marks on the fracture surface, formed during Stage II crack growth.

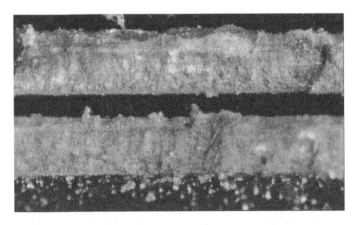

FIGURE 1.8
Optical fractograph of fatigue failure in 0.5 mm thick copper conductor laminations, showing the presence of faint, Stage II beach marks parallel to the plane of the Cu sheet.

evident markings on the surface such as ratchet marks and beach marks, which indicate the initiation point and direction of crack growth, respectively (Figures 1.7 and 1.8). Under high magnification, microscopic marks caused by the crack growth process, and known as *striations*, can be detected on the crack surface by SEM (Figure 1.9). Each striation would normally indicate one cycle of crack growth and striation counting would allow parameters to be fitted into the Paris [16] crack growth relationship (or later modifications)

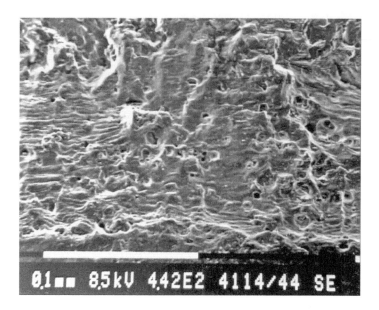

FIGURE 1.9
SEM fractograph of the copper sheet from Figure 1.8, showing striations arising from Stage II fatigue crack propagation.

to allow the Stage II cracking process to be modelled. Failure finally occurs by Stage III crack propagation when the remaining area becomes insufficient to withstand the applied stresses; fracture would be brittle if conditions were plane strain and ductile if plane stress.

1.4.3 Electrical Failure

The existence of a stable electrical supply is necessary for the functioning of a modern society but all too often power interruptions occur, which cause disruption of industrial, commercial, and day-to-day activities. Widespread 'blackouts' have occurred practically all over the world at one time or other, resulting in heavy financial losses and safety issues when airports, transportation, and health services become dysfunctional. Such disruptions are usually caused by some failure in the generation, transmission, or distribution system. Failures in the latter two are more frequent and are normally electrical in nature. Hence, there is a need to understand more about electrical failures, but texts in this discipline are quite sparse.

The transmission and distribution of electricity makes use of enormous amounts of conductors, transformers, and switchgear, most of which carry high voltages and currents. Local voltage levels generally range from 11 to 500 kV but levels of 1000 kV are also used in certain countries. The switchgear are subjected to particularly severe duty and suffer from a higher incidence of failures, the reason being that they are used to make and break heavy currents and voltages. The arcs arising from these actions must be quickly extinguished or damage will result; the extinguishing medium may be gas (SF_6), liquid (oil), or vacuum, all of which have high-dielectric strengths. In contrast, household switches are relatively safe devices because the low voltage levels only create small arcs that are easily extinguished in air.

There are two main electrical failure mechanisms, disruptive discharge (DD) across an insulating medium or overheating in a conductor by ohmic heat, as a result of current flow. An electrical conductor under voltage creates an electric field around it (MV/m); in normal circumstances, this field is carefully designed to be constrained by an electrical insulation system, but if for some reason the dielectric strength of the insulation is breached, an arc jumps from the conductor to the nearest electrode, which could be earth (ground) or another conductor of a different phase. A related mechanism is possible where current tracks across the surface of a contaminated solid insulator until the remaining insulated distance to another electrode is small enough for an arc to jump over. A DD can expend large amounts of energy and generate large amounts of plasma with temperatures of thousands of degree Celsius, often resulting in an explosion and fire. There would be several safety devices to cut off power within one or two seconds of such an occurrence, but if the devices were not healthy and if the discharge were sustained, very heavy damage would be caused to the equipment (Figure 1.10).

When current flows in a conductor, heat is generated in accordance with Ohm's law ($H = I^2R$), and in order to prevent overheating in electrical circuits,

FIGURE 1.10
(See Colour Insert.) Flashover at an electrical substation, compounded with protection failure, leading to a raging fire and total loss of the electrical switchgear.

FIGURE 1.11
(See Colour Insert.) Ohmic heat damage at a loose electrical joint, leading to melting, joint separation, arcing, and total failure.

conductors are made of metals with high electrical conductivity such as copper (Cu) and aluminum (Al). As long as they are properly designed and used under appropriate current densities, there is little danger of overheating, however, conductors need to be joined by bolting or spring-loaded contacts, especially in the vicinity of substations. When the joints become degraded, the electrical resistance rises and the ohmic heat that is generated may be sufficient to melt the contacting surfaces, creating open circuits and subsequent DD (Figure 1.11). A good proportion of substation faults arise from this mechanism.

1.4.4 Corrosion

Corrosion and its prevention is a very wide topic [17] so only a brief outline and some examples will be treated here. A good understanding of this topic is important because the effects of corrosion and/or its prevention costs hundreds of billions of U.S. dollars every year worldwide, for minimization, replacement, or repair. In general, all engineering metals and alloys would be susceptible to some form of corrosion or other and to prevent such occurrences would require a continuous financial outlay but to ignore them could cause greater losses. Gold, a noble metal, is practically corrosion resistant, but it is very expensive and its engineering usage is limited mainly to electrical contacts. On the other hand, stainless steels and Al alloys are highly reactive materials, but they form a very stable oxide film that protects against further attack. However, if localized breaks occur in the oxide film, which cannot be re-established, localized attack such as stress corrosion cracking, pitting, or intergranular cracking may occur. Corrosion wastage reduces the capacity of a component and if undetected, may result in some catastrophic failure. One noteworthy example is the Guadalajara Sewer Explosion in Mexico in 1992, which claimed more than 200 lives and damaged more than 1000 buildings [18]. Non-metals also suffer from corrosion, but their treatment is beyond the scope of this text.

Corrosion is the wastage or degradation of a material through chemical attack by environmental species, the most common of which are oxygen (oxidant) and water (electrolyte). In the absence of an electrolyte, and especially at elevated temperatures, oxidation takes place, resulting in the formation of oxide films. This process is often called *dry corrosion* and such corrosion is not an issue except in components such as boiler tubes or gas turbines, which are subjected to hot flue gases. Where both species are present together, electrochemical reactions are responsible and the process is often called *wet corrosion*. In this process, sites of different electrochemical potentials on the surface form a corrosion cell. At the more electronegative sites (anodic), the metal will go into solution, a wasting process; at the more electropositive sites (cathodic), several types of reactions may occur but no wastage takes place. The anodic and cathodic reactions for steel in oxygenated water may be written as:

At the anode,

$$Fe(s) \rightarrow Fe^{2+}(aq) + 2e^-$$ (1.2)

At the cathode, the steps can be any of

$$O_2 + 2H_2O + 4e^- \rightarrow 4OH^-$$ (1.3a)

$$H^+ + e^- \rightarrow \frac{1}{2}H_2(g)$$ (1.3b)

$$M^{2+} + 2e^- \rightarrow M(s)$$ (1.3c)

where

M is a metal.

The hydroxide formed in Equation (1.3a) quickly oxidizes to form rust. On corroding steel surfaces, the anodic and cathodic sites will shift around and after some time the whole surface would be attacked and covered with rust. This form of corrosion is called *uniform corrosion* and is not normally a dangerous form as the attack is easily detected. More dangerous ones arise from some form of localized attack where a small anode is linked to a large cathode; attack can be very fast and often results in unexpected failures through unawareness or difficulty in detection. Some localized forms include crevice corrosion, pitting, intergranular granular attack, corrosion under insulation, microbial corrosion, and stress corrosion cracking. Following are some interesting examples of the last three types of corrosion.

1.4.4.1 Corrosion under Insulation

Corrosion under insulation (CUI) occurs on surfaces that are wrapped and covered up with thermal insulation such as rock wool or glass wool, for example on the exterior of steam pipelines in power plants or in the space between double-layered roofs of large buildings. If the external cover is not watertight, rainwater can seep in and provide regular infusions of fresh, oxygenated water. The situation then becomes that of an *open system*, where polarization processes cannot take place to stifle the corrosion processes, as would occur in a *closed system*. Wastage can be rapid and the only possible solution is to have a fully watertight cover and/or to use a good barrier coating such as a two-part epoxy. Galvanizing does not work under such conditions. Figure 1.12 shows an example of CUI in the galvanized steel spacers supporting the stainless steel skin of a very large roof. Rainwater that had entered during construction had wet the thermal insulation and had not dried before the skin was installed. Fortunately, this was realized by the owner and remedial measures were applied before the building was commissioned, otherwise some very expensive and inconvenient repairs would have had to be made at a later time (Chapter 7, Case Study 1).

FIGURE 1.12
(See Colour Insert.) Corrosion under insulation (CUI) of galvanized mild steel spacers supporting the outer, stainless steel skin of a very large, rock wool insulated roof, caused by rainwater entry during installation.

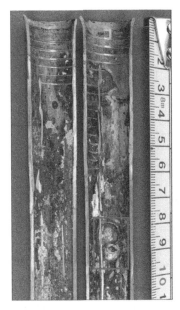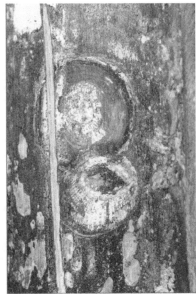

FIGURE 1.13
(See Colour Insert.) Microbial corrosion beneath barnacle bases in an aluminum brass, heat exchanger tube with seawater flowing inside; the boxed area at the left figure is magnified at the inset on the right.

1.4.4.2 Microbial Corrosion

Microbiologically influenced corrosion (MIC) or *microbial corrosion* for short, is essentially an attack promoted by the action of bacteria which then changes the environment of the fluid in contact with the metal; sulfate reducing bacteria being a typical example. The fluid does not have to be water and MIC has often been found to occur at the bottom of oil storage tanks and oil pipelines. Figure 1.13 illustrates MIC at the internal surfaces of Al brass heat exchanger tubes in a power plant. The seawater going into the tubes is normally dosed with sufficient chlorine to kill off all the sea life, but there was a malfunction in the dosing system during commissioning tests and further, the seawater was not drained off after the event, a double fault situation. Consequently, colonies of barnacles and bacteria had been able to establish themselves; the bacteria caused the tube metal to be attacked and during operation several months later many tubes were found to be punctured. The presence of the barnacles was incidental and had only helped the bacteria to flourish.

1.4.4.3 Stress Corrosion Cracking

Stress corrosion cracking (SCC) is a commonly encountered form of attack in metals that depends upon a stable oxide layer for protection, such as stainless steels and aluminum alloys. However, it is quite rare in carbon steels

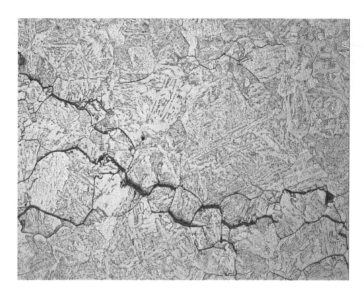

FIGURE 1.14
Microstructure of nitrate stress corrosion cracking of a large, carbon steel, nitrogen vessel, showing intergranular cracks in the ferrite/bainite microstructure at the weld joint. Mag. 500×.

except in environments containing nitrates, carbon dioxide, and certain alkalies. Figure 1.14 shows a case where a large, long-serving nitrogen vessel had exploded due to nitrate SCC at the weld joints. The nitrogen was obtained by combustion of natural gas with atmospheric air and then removing the byproduct gases. The presence of the nitrates was puzzling, but it is possible that the atmosphere became contaminated with compounds of nitrogen from an ammonia plant next door, became converted to oxides of nitrogen during combustion, and eventually formed nitrates within the vessel through reaction with water vapor from the air (which was periodically allowed to enter). The vessel had undergone the mandatory annual inspections before it was allowed to be used, but inspection was only limited to thickness checks and integrity of the paint coating; non-destructive examination (NDE) of the welds for the presence of cracks was not performed. A second, unexploded vessel was subsequently examined and scrapped, as it was also found to contain extensive cracks.

1.5 Prevention of Failures

The various stages in the engineering process that could result in failures are

1. Design and material selection,
2. Manufacturing, installation, and quality control, and
3. Operation and maintenance.

1.5.1 Design and Material Selection

The designing of a component involves the identification of all the stressors operating and the calculation of the resultant stresses. Examples of methodologies used for the calculation of mechanical stresses are provided in Chapter 2. These methodologies as well as those used for structural design and electrical design have been incorporated into codes of practice (CP) or standards. Different countries have their own codes but their methodologies do not vary significantly; a professional design engineer is mandated by law to follow the codes recognized by the authorities of the country in which he or she practices. If these codes are adhered to, the designed component is expected to be able to function as expected, without premature failure. Errors may sometimes be present in new codes that have not withstood rigorous usage, but such errors have not been found to be a serious source of failures and are usually detected early in the life of the code. On the other hand, errors in the interpretation of the codes due to incompetence are more prevalent and do occasionally cause failures. For this reason, engineering design should only be undertaken, or at least closely supervised, by experienced professional engineers. It must be cautioned though, that good design analysis needs to be complemented by a good appreciation of the nature of materials and the methods to obtain properties in built components, which can equal the levels obtained in laboratory tests of standard specimens.

The incorrect specifying and manufacture of materials of construction are larger sources of failures than poor design analysis and this responsibility also falls within the purview of the designer. The materials used in engineering construction/equipment and requiring structural strength are extremely wide, but may be divided into non-metallic and metallic materials. The non-metallic materials have concrete and plastics as the main sub-groups. The metallic materials may be divided into ferrous and non-ferrous sub-groups. The ferrous sub-group encompasses all the various types of steels and cast irons. The non-ferrous sub-group includes all other engineering metals and alloys, derived variously from aluminum, copper, titanium, magnesium, zinc, and other more exotic alloys for specialized usage. The following sections will provide brief details of the two most commonly used sub-groups, namely, concrete and steels.

1.5.1.1 Concrete

Concrete has been used for many centuries and most of the ancient civilizations built some form of concrete structure or other, which has survived to the present age. This would give the impression that concrete is a simple material but that is a fallacy. Present day materials science has shown that it has a very complex structure and that its various properties can be modified by proper mixture control and processing to give the necessary microstructures. For instance, in the 1950s and 1960s, it was quite normal to use concretes with a design strength of about 20 MPa, but today, strengths in excess of 100 MPa [19] are quite commonplace, especially for high-rise structures. An important characteristic of concrete is that it is strong in compression, but weak in tension, and engineers have overcome this inadequacy by introducing steel reinforcement bars (rebar) at locations of tensile stress. In an extension of this principle, the use of prestressing has allowed higher loads than ever before to be borne. In this method, areas subjected to tensile stress during operation are given an initial compressive stress by the use of high-strength steel wires, which will counteract the operational tensile stress.

A concrete can be looked upon as a *glue* mixture (binder) with inert fillers such as natural aggregates (sand, limestone, or crushed granite), man-made fibers, and steel rebar. The glue would comprise a cementitious material, water, and air; the most common cementitious material would be a hydraulic cement such as Portland cement, which, when mixed with water will undergo a hydration process. Initially, the mixture will become plastic and gel-like, during which time the whole mass of glue with fillers can be worked and shaped (workability). Eventually, chemical reactions will ensue (hydration) and the glue will harden to give a solid mass (curing). The hardening process should give sufficient strength after a certain period to allow the formwork to be removed. Concretes are often specified by their 3-day, or 7-day, or 28-day strengths, though the hardening process proceeds for a much longer time than 28 days. Concretes have the respective approximate compositions [20]: cement (9 to 15%), water (15 to 16%), fine aggregates (25 to 35%), and coarse aggregates (30 to 45%). Cement and water constitute the paste (glue); the paste and fine aggregates constitute the mortar, and the mortar and coarse aggregates constitute the concrete. The rationale for using both fine and coarse aggregates is that when mixed together with the cement paste, the smaller particles will fill the voids between the coarse aggregates to maximize aggregate content.

Concretes, including the reinforced and prestressed forms, are the main construction materials used by civil/structural engineers and they have generally performed very well after passing through the construction stage, but occasionally, failures or material distress occur, which may take the form of total collapse, or cracking, or degradation of the concrete or reinforcement [21]. In properly specified material, root causes arising from inadequate material properties can be prevented through the setting up of a quality assurance system by the owner, which, amongst other functions, is able to audit the quality control being carried out by the contractor.

1.5.1.2 Steels

Steels are a very versatile and important class of engineering material and without them the world today would be devoid of many technological achievements, such as aircraft, bullet trains, high-tonnage ships, and even ordinary motor vehicles. Their wide applications can be illustrated by just a brief review of the most commonly used groups:

1. Weldable structural steels with ferritic/pearlitic microstructures, with C up to about 0.25% and Mn up to about 1.6%; used for ships, bridges, and general engineering.

2. High-strength steels for mechanical components, with a tempered martensite microstructure, containing C up to about 0.4% and variable amounts of alloying additions such as Cr, Mo, and Ni. Such steels require careful processing and heat treatment and are used for shafts, spindles, gears, bolting, aircraft landing gears, and similar highly stressed components. This group includes the high-performance maraging steels.

3. Hard steels with a lightly tempered martensite microstructure containing high volumes of hard precipitates; high-carbon and high-alloying additions; used for tools, dies, and moulds.

4. Stainless steels with austenitic, ferritic, or martensitic microstructures depending upon alloying additions and heat treatment. They have Cr greater than 12%, with varying amounts of C, Ni, Mo, and other alloys and are used for a wide variety of applications requiring resistance to the environment and the various chemicals present in processing plants. The precipitation hardening grades of martensitic and austenitic steels require special heat treatment processes.

5. High-temperature steels with ferritic/pearlitic, martensitic or austenitic microstructures, with and without stable precipitate forms. They have variable alloying additions; used for boiler components, steam, and gas turbines.

The essential properties required, though not all at the same time, include tensile and compressive strength, ductility and toughness, fatigue strength, hardness and abrasion resistance, corrosion resistance, high-temperature creep resistance, and low-temperature toughness; any significant inadequacies in these properties may result in eventual failure. Certain failure modes are common to all the groups but some will be specific to the group itself. For instance, the failure modes for a structural steel would be fracture, fatigue or corrosion; for a hard tool steel the modes would be fracture and/or wear; for a boiler steel the modes would be fracture, creep, thermal fatigue, or corrosion. For a high-strength steel, brittle fracture and fatigue failure would be the main mechanisms. The prevention of failure of steel components from material inadequacy for any particular type of application requires that the appropriate properties be achieved. The respective properties are obtained by proper alloying, thermo-mechanical processing and heat treatment. Unlike concrete manufacture, where small localized mixing plants are able to produce to the quality required, steel manufacture needs economies of scale, which in turn requires the outlay of heavy investments to be profitable. This has resulted in steels with apparently similar specifications but having unequal performance. On paper, a steel from a highly reputable manufacturer with new steel-making facilities or from a third world country with ancient facilities may appear to conform to the required grade, but in usage, the steel by the latter may fail due to an accumulation of small inadequacies that are not normally tested for in the acceptance standards. For example, a weldable structural steel with a coarse grain size and with C, Mn, S, and P contents at the upper acceptance limits, and with a high-impurity content (Sn, As, and Sb), would be a good candidate for BF. Where a project requires large quantities of steels to be bought from an unproven source, it would be prudent to have the source audited by a competent person and to adopt a stringent quality assurance (QA) system.

1.5.2 Manufacturing, Installation, and Quality Control

1.5.2.1 Concrete

Concrete needs to be poured into a properly supported formwork and between the moment the constituents are mixed with water and the time it starts to set, the mix must be properly compacted. Two parameters therefore need to be controlled here, the setting time and the pouring/compaction time. The setting time must be sufficient to allow proper pouring/compaction but not so excessive as to prolong hardening to long periods of time. This time can be fairly accurately controlled by concrete technologists by varying the nature of the mix and by the use of chemical admixtures. Pouring and compaction at site need to be ensured by having an efficient work flow. If any batch of concrete does not comfortably meet the scheduled tests before pouring it would be prudent to discard it, because bad concrete entering the formwork would take a lot of time and trouble to rectify.

A matter not related to concrete behavior is the improper detailing of rebar at T-joints and corners, which results in weak joints that are prone to cracking; this fault can be attributed to poor construction practice and supervision [21]. There are various other inadequacies that can occur, but all these can be minimized by setting up a proper QA/QC system and employing competent people who understand the properties of the materials used and who are skilled in modern construction practice. Case Study 2 in Chapter 7 gives an example of poor construction practice that had to be rectified at a high cost.

1.5.2.2 Steel

The types of components made of the various types of steels are too many to describe, even very briefly, but a frequent location of failures, welded joints, will be mentioned here. In modern joining practice, welding has almost totally replaced riveting and bolting. Welding can be simply considered as the creation of a molten pool to fuse two adjoining parts, and then allowing the joint to cool down to make an integral component. The parts to be joined may be made of similar or dissimilar materials of similar or dissimilar thicknesses. Most metals can be joined by welding but the crux of the matter is whether the properties and serviceability of the joint will equal that of a similar component made of a single piece of material. Hence, some questions need to be answered before any weld joint is made, such as:

Q1: What are the material types and thicknesses that can be safely welded together?

Q2: What weld metal and welding procedures need to be used?

Q3: How can the metallurgical quality of the joint be assured?

Q4: How can the resulting distortions and residual stresses be kept to acceptable levels?

Q5: What are the skills required to make a sound joint?

Q6: What are the performance limits of the joint?

These questions have been closely considered by players in the various steel fabrication industries the world over. They include experts involved in structural steel (cranes, bridges, ships), chemical and petrochemical plants (stainless steel process vessels and pipelines; boiler components), and power plants (high-temperature boiler components, pressure vessels, and pipelines). They have formulated safe practices that have been incorporated into various codes of practice and standards, which differ from country to country in only minor aspects. All codes will attempt to ensure reliability and quality of welding by requiring code users to perform welding procedure tests, and welders to undergo periodic welder qualification tests. Depending upon the materials used and the joint configuration, various levels of preheat treatment and post weld heat treatment (PWHT) may be applicable. Furthermore, welded joints need to undergo some form of non-destructive evaluation (NDE) to quantify surface defects (e.g., by magnetic particle, penetrant dye, and eddy current) and volumetric defects (e.g., by radiography or ultrasonic testing). Often, to ensure that PWHT has been properly performed to soften the hard heat affected zones in transformable steels (e.g., ASME SA-213 Grade T91), *in situ* hardness tests need to be performed using portable hardness testers and *in situ* metallography may need to be performed via plastic replicas. If all the required procedures are strictly followed, weld joints can attain high levels of serviceability, though their fatigue strengths will never equal that of an unwelded component; if this issue is critical, the part can be suitably re-designed.

1.5.3 Operation and Maintenance

When these terms are used they are immediately associated with machinery, which have moving parts and more often than not need some form of lubrication; the motorcar is a common example. However, even static concrete structures have to be properly operated and maintained; for example, concrete structures are designed with a maximum expected, continuous loading, and if this limit is often exceeded, the serviceability will be reduced and failure may eventually ensue. Periodic inspection is also required to ensure that drains and culverts are not blocked, that the cover has not spalled off to expose rebar, that there has been no sulfate attack in marine atmospheres, that protective paintwork is intact, or that cracks are not present at joints or areas of tensile stress. Where machinery is concerned the operation and maintenance guidelines need to be closely adhered to, in order to detect the presence of wear and tear, corrosion, high-temperature degradation, fatigue cracking, deformation, or other mechanisms that will reduce serviceability and safety. Maintenance should be planned and should not be reactive.

Problems and Answers

Problem 1.1

A round bar made of mild steel, broken in a tensile test, exhibited a cup and cone failure. Examination showed that the central portion was perpendicular to the bar axis but that the sides were inclined at about 45° to bar axis. (a) What would their fractographic appearances be, at 2× and 2000× magnification? (b) How can these magnifications be obtained?

Answer 1.1

(a) The central portion, at 2× magnification, would have a gray, fibrous appearance and at 2000× magnification, would show equiaxed dimples. The sides, at 2× magnification, would have a smooth silky appearance and at 2000× magnification, would show elongated dimples.

(b) A 2× magnification can be obtained via a simple magnifying glass. A 2000× magnification can only be obtained by use of an electron microscope.

Problem 1.2

An infinite steel plate, 30 mm thick, is subjected to a tensile stress. It contains a through thickness crack perpendicular to the stress direction. (a) Calculate the fracture stress (σ_c) for a crack length ($2a$) of 50 mm, assuming plane strain conditions. (b) If conditions were plane stress instead, what would be the failure condition, and at what expected stress level? (c) What microstructure would you expect such a steel to possess?

Given

$$K_{IC} = 80 \text{ MPa m}^{-1/2}; \text{ yield stress } (S_y) = 1000 \text{ MPa.}$$

Assume $Y = 1$ for an infinite plate.

Answer 1.2

(a) Work in m, MPa, and MPa m$^{-1/2}$ units. Use Equation (1.1):

$$\sigma_c = \frac{K_{IC}}{Y (\pi a_c)^{1/2}} = 80/(\pi \times 0.025)^{1/2} = 285.5 \text{ MPa}$$

Comment: The brittle fracture stress under plane strain conditions is much lower than the yield stress of the material.

(b) Under plane stress conditions, ductile fracture would occur and the material would first yield at a stress of 1000 MPa. For this to occur, the plate thickness and crack length need to be less than the value of $2.5 \left\{ \dfrac{K_{IC}}{S_y} \right\}^2$ given in Equation (2.1), though this must not be regarded as a strict 'go-no go' type of condition. The condition can be satisfied by increasing K_{IC} or reducing S_y or both. It is a general rule that in high-strength steels, the toughness is inversely proportional to strength.

(c) The steel would very probably have a tempered martensite micro-structure, obtained through appropriate alloying and heat treatment.

Problem 1.3

In Figure 1.4, give three possible reasons why the punched hole had initiated brittle fracture.

Answer 1.3

(a) The hole would create a fairly high stress concentration that would add to the applied stress.

(b) A layer of material surrounding the bore of the hole would be heavily deformed and consequently would have its ductility fully exhausted; it would be quite brittle and probably contain small cracks, which could extend under stress.

(c) The deformed layer would also contain high residual stresses that would add to the total stress.

Problem 1.4

Why is it important to have adequate fillets in machinery components where the section sizes are variable?

Answer 1.4

Machinery is invariably subjected to impact loading and variable loading. Sharp changes in section induce stress concentrations of high magnitudes that promote brittle fracture under impact loads and fatigue cracking under variable loading.

Problem 1.5

A 2000 W (P) electric kettle is powered by a 240 V, 3-phase supply, via a plug with a fuse of 13 A. (a) Calculate the current (I) that it will draw and the resistance (R) of the heating elements. (b) If for some reason the resistance suddenly falls to 2 Ω during use, what would be the consequences?

Answer 1.5

Ohm's law applies. For 3-phase current, the current (I) and resistance (Ω) are:

(a) $I = P/(V \times 1.732) = 4.81$ A; $R = (V \times 1.732)/I = 86.42$ Ω
(b) If R falls to 2 Ω, $I = (V \times 1.732)/2 = 207.84$ A. Since the fuse is rated at 13 A, it will blow and cut off the current to the kettle. (A blown fuse is symptomatic of a low resistance path caused by conductors shorting to earth [ground]; it would be wise to investigate before reconnecting power. Own reading—what is the factor 1.732?)

Problem 1.6

(a) Name a code of practice for welding steels that is commonly used the world over. (b) PWHT is not used for austenitic steels. Why? (c) In your opinion, why were the World War II Liberty ships prone to failure by brittle fracture, whereas the older riveted ships were not similarly affected?

Answer 1.6

(a) American Welding Society (AWS) Structural Welding Codes—Steel (D1.1/D1.1M).
(b) Austenitic steels do not transform to hard microstructures on welding and do not need to be softened. Conversely, such treatments will cause grain boundary carbides to precipitate and sensitize the steel to intergranular corrosion.

(c) The residual stresses from welding helped to initiate brittle fracture; the steels used at that time were inherently brittle (high transition temperature) and allowed fracture to propagate once initiated. Further, the ships had hulls welded into a continuous component and the brittle crack could traverse from one side to the other without restraint, to break the ship into half.

On the other hand, riveted ships had hulls composed of plate sections; the riveting did not have high residual stresses to promote brittle fracture initiation and any fracture that a plate experienced would be confined to that plate alone.

References

General References

ASM Handbook, Vol. 9, *Metallography and Microstructures*, Ed., G.F. Vander Voort, ASM International, Ohio, 2004.

ASM Handbook, Vol. 11, *Failure Analysis and Prevention*, Ed., W.T. Becker and R.J. Shipley, ASM International, Ohio, 2002.

ASM Handbook, Vol. 12, *Fractography*, Ed. R.L. Stedfeld et al., ASM International, Ohio, 1987.

ASM Handbook, Vol. 13A, *Corrosion: Fundamentals, Testing, and Protection*, Ed., S.D. Cramer and B.S. Covino, Jr., ASM International, Ohio, 2003.

ASM Handbook, Vol. 19, *Fatigue and Fracture*, S.R. Lampman et al., ASM International, Ohio, 1996.

Specific References

1. *Toyota USA Newsroom Amended Recall: Potential Floor Mat Interference with Accelerator Pedal*, Toyota Corporation, 2009–2011. http://toyota.tekgroupweb.com/safety-recall/. Accessed May 2012.
2. Takahashi, Y., Toyota: 400,000 Vehicles Subject to Recall Worldwide, *Wall Street Journal*. http://online.wsj.com/article/BT-CO-20100209-702754.html?mod=WSJ_World_MIDDLEHeadlinesAsia. Accessed 5 May 2012.
3. Haq, H., Toyota Recall Update: Dealers Face Full Lots, Anxious Customers, *The Christian Science Monitor*, 1 January 2010. http://www.csmonitor.com/USA/2010/0129/Toyota-recall-update-dealers-face-full-lots-anxious-customers. Accessed 8 May 2012.
4. Radiation-Exposed Workers to Be Treated at Chiba Hospital, *Kyodo News*, 25 March 2011. http://english.kyodonews.jp/news/2011/03/80994.html. Accessed 8 May 2012.
5. Japan's Unfolding Disaster Bigger than Chernobyl, *The New Zealand Herald*, 2 April 2011. http://www.nzherald.co.nz/world/news/article.cfm?c_id = 2&objectid = 10716671. Accessed 8 May 2012.

6. Wakatsuki, Y. and K. Lah, CNN, 3 Nuclear Reactors Melted Down after Quake, Japan Confirms, *CNN*, 7 June 2011. edition.cnn.com/2011/WORLD/asiapcf/06/06/japan.nuclear.meltdown/index.html?iref = NS1. Accessed 8 May 2012.

7. *BS 7608: 1993: Code of Practice for Fatigue Design and Assessment of Steel Structures.* British Standards Institution, London, 1993.

8. *BS EN 1993-1-9: 2005, Eurocode 3, Design of Steel Structures. Fatigue.* British Standards Institution, London, 2005.

9. *ASTM E1820: Standard Test Method for Measurement of Fracture Toughness*, ASTM International, West Conshohocken, PA, 2013.

10. Bowles, C.Q., *Fracture and Structure*, in ASM Handbook, Vol. 19, *Fatigue and Fracture*, S.R. Lampman et al., ASM International, Ohio, 1996.

11. Kerlins, V., *Modes of Fracture*, in ASM Handbook, Vol. 12, *Fractography*, R.L. Stedfeld et al., ASM International, Ohio, 1987.

12. Rosenfield, A.R., *Fracture Mechanics in Failure Analysis*, in ASM Handbook, Vol. 19, *Fatigue and Fracture*, S.R. Lampman et al., ASM International, Ohio, 1996.

13. Irwin, G.R., *Fracture Dynamics*, Trans. ASM, Vol. 40A, 1948, pp. 147–166.

14. Broek, D., *The Practical Use of Fracture Mechanics*, Dordrecht, Boston, London: Kluwer Academic, 1989, pp. 433–434.

15. *Fatigue Failures*, in ASM Handbook, Vol. 11, *Failure Analysis and Prevention*, Ed., W.T. Becker and R.J. Shipley, ASM International, Ohio, 2003, pp. 1469–1516.

16. Paris, P.C. and F. Erdogan, A Critical Analysis of Crack Propagation Laws, *J. Basic Eng.*, Trans. ASME, Vol. 85, 1963, pp. 528–534.

17. Covino, Jr., C.S. and S.D. Cramer, *Introduction to Forms of Corrosion*, in ASM Handbook, Vol. 13A, *Corrosion: Fundamentals, Testing, and Protection*, Ed., S.D. Cramer and B.S. Covino, Jr., ASM International, Ohio, 2003, p. 189.

18. Sewers Explode in Guadalajara, *The History Channel*, 2012. http://www.history.com/this-day-in-history/sewers-explode-in-guadalajara. Accessed 10 May 2012.

19. Day, K.W., *Concrete Mix Design, Quality Control and Specification*, 3rd ed., London and New York: Taylor & Francis, 2006, p. 12.

20. Taylor, P.C., S.H. Kosmatka, G.F. Voigt et al., *Integrated Materials and Construction Practices for Concrete Pavement: A State-of-the-Art Practice Manual*, FHWA Publication No. HIF-07-004, U.S. Department of Transportation, Federal Highway Administration, Washington DC, December 2006.

21. Whittle, R., *Failures in Concrete Structures*, Boca Raton, FL: CRC Press, 2012.

2

Failure Analysis Procedures

2.1 General Introduction

The techniques used in the investigation of engineering failures are wide ranging and are continuously expanding as more and more sophisticated instruments, equipment, and computing software become invented and made affordable. These are used, for example, for the measurement of miniscule amounts of elements, or for the examination of sub-microscopic phases or surfaces of materials, or for the 3-dimensional modeling of stresses and deformations of complex objects. All these have allowed the Failure Investigator (FI) to identify the proximate and root causes of a failure more accurately than ever before. At the same time, parallel developments in other fields have enabled the FI to produce reports faster. For example, with the present developments in digital photography and printing, computer hardware and software, and Internet technology, the FI is able conduct site examination, process the relevant photographs, look up some online literature, write a short preliminary report, and have a hard copy ready by the next day, or have an electronic copy sent by Internet halfway round the world in an even shorter space of time. Two decades ago, such speed and efficiency was quite inconceivable. But, despite all of the above advancements, a carefully planned procedure of investigation, directed by human skills and performed by human investigators, must first be conducted before any findings can be made or any report can be produced. Such a procedure should include some or all of the following steps:

1. Obtaining background information
2. Physical examination/testing and chemical analysis
3. Stress analysis and computer modeling
4. Report writing

In the process of conducting failure analysis (FA), the FI has to look beyond the normal limits of design methodology, and identify the actual conditions causing the failure, and sometimes check the design. This requires familiarity with the major types of stresses and strains and failure modes, disciplines that can normally only be obtained through a rigorous academic program, experience on the field, and self-study. The FI also needs to be recognized as being professionally competent by the regulations in force, and to be deemed an expert witness, whose findings can be accepted in court. The FI is required to be a technical detective as well as a student of human psychology, for the simple reason that persons who have been negligent and have caused the failure do not normally incriminate themselves by telling the truth. The FI should be wise enough to detect an untruthful witness.

2.2 Obtaining Background Information

This is a crucial step in the procedure, as no FI faced with an unfamiliar case would be able to conduct a satisfactory investigation without knowing the background, which would include the relevant parts of the history of the failed component. However, more likely than not, especially where the failure concerns old components, the early history (design, fabrication, operation, and maintenance) would be incomplete and the ingenuity of the FI would then be sorely tested. A site visit to inspect the scene of the failure is necessary to obtain first-hand information, except for simple cases. But before doing so it would be useful to first obtain a short account of the incident and as many photographs as possible, as this would allow the FI to have preliminary thoughts about the failure and to conduct a fast literature search. Relevant historical information can then be asked for, which can be viewed at the site or at a later date. The steps to take in obtaining the required information are outlined below.

2.2.1 During Site Inspection

Site inspections are necessary to obtain first-hand information concerning the nature of failure, and to be effective the FI must be properly prepared for such a task. They must firstly be equipped with personal safety equipment, helmets, safety shoes, and gloves at the minimum. For visual inspections, it would be necessary to have a good digital camera (or two), measuring instruments (a measuring tape, a pair of 30 cm calipers, and a 30 cm steel rule), a hand-held magnifying lens, a small mirror to look around corners, a small powerful torchlight, writing materials and/or voice recorder, thick plastic bags for small specimens, some plasticine, some rags and hand cleaner; all these can fit into a small bag. Assistant investigators could help to perform the menial tasks but should not be fully entrusted with obtaining the photographic evidence unless they have been fully trained to do so. The reason is that the initial evidence must be properly recorded, a task demanding intimate knowledge of

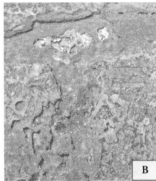

FIGURE 2.1
(See Colour Insert.) (A) Site photographs of CO_2 attack on API 5CT, Grade L-80, production tubes from an offshore oil well. (B) Shows a close-up view of a typical attacked location with darkish $FeCO_3$.

failure modes and good photographic skills, which only come with training and experience. The site inspection should be conducted as soon as possible before evidence becomes tainted and if possible, the site should be cordoned off until investigations are completed. There is often a pressing need to allow repairs to be made so that operations can resume as soon as possible; hence site work must be performed quickly and efficiently.

Appropriate notes and photographs should be taken at all stages of site inspection. When taking photographs, initially take overall views from different angles to show the relative locations of the different components, and then zoom in progressively to the location of failure initiation for details of the failure characteristics. A lens with macro capabilities would be an advantage (Figure 2.1). The work scope during the site inspection generally includes some or all of the tasks below, though in complex cases the list can be longer.

1. *Evaluating the functional and operational aspects of the failed component.* A briefing from the owner or operator is essential for understanding complex machinery; if other similar components or machines are present, they should be used as examples.

2. *Identifying the nature of the failure and collecting of samples.* The overall scene should be inspected to identify the part that failed first. Parts without plastic deformation suggest failure below yield stress, a good indicator of brittle or fatigue failures. Such parts will often have a fracture origin with a pre-existing crack, which is usually stained or corroded as a result of exposure to the environment (Section 2.3). Where there has been plastic collapse of a multi-element structure, the pattern of collapse should be identified to enable the initiation point to be accurately located.

Loose samples (e.g., broken pieces, corrosion scale, lubricating oils, or other liquids) that can be easily carried should be acquired after their location and orientation have been carefully photographed. Larger samples that need to be cut out also have to be carefully photographed and then marked for extraction. The manner of extraction must be carefully spelt out, as careless flame cutting may cause unintended changes in microstructures and material properties.

3. *Getting witness accounts.* Eyewitnesses and personnel operating and maintaining the component should be interrogated and if necessary, be asked to sign written witness statements. Events leading to the occurrence of the failure, as well as recent occurrences of abnormal behavior should be fully recorded.

4. *Examining recorded data.* Data in chart recorders or in electronic recorders should be examined and copies made of the relevant records if required. Where non-recording gauges are used, their set-points should be recorded and compared with specifications. Where handwritten logbooks are used, they should be examined and copies made of relevant entries. It should be noted that a good digital camera is very handy for making copies at the site.

5. *Obtaining detailed operational and maintenance practices.* Current and past practices need to be obtained for comparison with specified procedures.

6. *Obtaining details of the manpower involved.* This is necessary to evaluate their adequacy and competency.

7. *Conducting* in situ *tests.* A wide range of tests can be conducted at the site to allow fast results to be obtained, without waiting for the necessary samples to be extracted and brought to the laboratory for testing. Further, some of the components may still be reusable and should not be cut up without a very good reason. Such tests include non-destructive testing (NDT), replica extraction for metallography, hardness testing for material properties, and positive material identification (PMI) using portable equipment (such as a spark-optical emission spectrometer and X-ray fluorescence).

2.2.2 Post-Site Inspection

Upon returning to base and after examining all the evidence obtained and conducting a preliminary examination of the samples collected, the FI should have a fair idea of what has occurred and what further information is required. This may include some or all of the following:

1. Design codes used and design specifications; operational and maintenance manuals; detailed drawings of the failed components.

2. Commissioning history; detailed operation and maintenance history, in particular, recent servicing and overhauls.

3. Written reports on the incident (by in-house personnel, the fire brigade, police, or insurance adjusters).

4. Recorded operational data at the time of commissioning and just after, also before and after major overhauls and for a reasonable period of time before the incident.

5. In most cases, a literature survey would be useful to look for current information; the Internet is a powerful search machine but care must be taken to ensure that references used are reputable.

6. The rest of the information needed to support the investigation would be obtained from laboratory work (Section 2.3).

2.3 Physical Examination/Testing and Chemical Analysis

The activities involved are wide ranging but those most frequently used in the analysis of metallic materials include:

1. Visual examination

2. Metallurgical examination (and quantification)

3. Mechanical testing and chemical analysis

4. Other tests and physical modeling

2.3.1 Visual Examination

A detailed visual examination in the laboratory should be performed first to reinforce observations made at the site, to determine whether further site visits are needed for a better understanding of the site conditions, or if more samples are required for testing. This involves an examination of the acquired samples in the as-received state and after cleaning up, trying to ascertain how and why they failed. Simple tools can be used at this stage that will not cause any loss in evidence, such as a magnifying glass or a portable hardness tester, or even a PMI instrument for material identification. Procedures would be directed towards evaluating or identifying the following:

1. The general quality of manufacture and whether the shape and dimensions of the samples conform to specifications.

2. The manner of deformation, whether caused before or during failure.

3. The presence of wastage and degradation, whether from normal wear and tear, abuse in service, corrosion, or erosion.

4. The fracture characteristics, whether caused by ductile, brittle, fatigue, or other mechanisms.

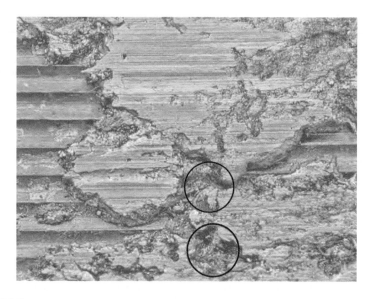

FIGURE 2.2
Macro-photograph of solidification shrinkage defects (circled) at the white metal layer of a bearing pad, causing premature failure. Mag. 8×.

  5. Locations of crack initiation where fracture/failure had been preceded by cracking.
  6. The appearance of the fracture surfaces, whether old or freshly formed.

At all stages, proper photographs should be taken and if any surface soils or debris are to be removed, samples should be taken, documented, and stored for later analysis, if required. High-quality technical photography is essential to show the relevant characteristics in detail because clear illustrations can make the difference between good and mediocre reports. Figure 2.2, which shows solidification defects at the white metal layer of a bearing pad, is a good example. The photograph was captured with a medium range, 16 Mp, APS-C digital camera with a macro-lens. It would be unwise to proceed beyond the visual examination stage and to start cutting up indiscriminately to obtain samples before the proximate causes have been identified to a fair degree of certainty.

2.3.2 Metallurgical Examination (and Quantification)

The objectives include establishing the modes of failure, the fracture mechanisms, the microstructural properties, and the presence of any production or metallurgical inadequacies that could have contributed to failure. The scope of examination is detailed in Table 2.1 and would normally encompass the following:

1. *Fractography*. This involves a study and quantification of the fracture surfaces under magnification, via stereo-microscopy or electron microscopy.
2. *Metallography*. This involves a study and quantification of prepared sectioned planes, with and without etching, under magnification, via optical or electron microscopy.

The above examination is performed to extend the observations made by visual examination, to smaller and smaller areas, by the use of instruments under higher and higher magnifications. The instruments used include optical ones with light as the signal carrier or electronic ones with electrons as the signal carrier. Visible light has wavelengths of the order of 390 to 700 nm and a maximum resolution of up to about 250 nm. The wavelengths of electrons are much smaller than that of visible light, and fall with accelerating voltage. At 10 kV, their wavelength would be about 12.2 pm, falling to about 2.5 pm at 200 kV. Consequently, the resolutions of electron microscopes are much higher than that of light microscopes and in modern day instruments, exceed 1 nm (Table 2.2). The instruments used also have attachments that can quantify certain mechanical properties such as phase hardness or metallurgical properties such as crystallographic orientation (Figure 2.3) [1], or elemental compositions. The ability to perform elemental analysis on microscopic phases or corrosion products on fracture surfaces is very useful. In the first instance, a concentration of a certain phase on the fracture surface could

TABLE 2.1

Modes of Metallurgical Examination

	Type of Examination	
Magnification	Fractography of Fracture Surfaces	Metallography of Prepared Sections
Macroscopic (1× to 50×)	Conducted on large pieces before and after cleaning, with a magnifying lens or stereo-microscope.	Conducted on large sections with a magnifying lens or stereo-microscope.
Microscopic (25× to 1500×)	Conducted on small pieces with a stereo-microscope up to 100×, or SEM/TEM to higher magnifications; a TEM needs a replica of the surface.	Conducted on small sections, most often with an optical microscope, in special cases with a TEM or SEM. A TEM requires a thin section to be prepared or a replica to be made. The thin section gives the bulk microstructure.
Sub-microscopic (1500× to > 1,000,000×)	Conducted on small pieces with a SEM/TEM; a TEM needs a replica of the surface.	Conducted on small sections with a TEM or SEM. A TEM requires a thin section to be prepared or a replica to be made. The thin section gives the bulk microstructure.

TABLE 2.2

Typical Features of Optical and Electron Microscopes Used for Failure Analysis

	Instrument		
Basic Features	**Light Microscope**	**SEM**	**TEM**
Illumination or excitation method	Visible incident light: Focused beam in a metallographic microscope. Non-focused beam in a stereo-microscope.	A focused electron beam for performing a raster scan of an area on the surface.	A focused electron beam for penetrating a thin film, typically <100 nm thick.
Power source/ accelerating voltage	Light source to 100 W typical.	0.3 to 30 kV typical.	40 to 200 kV typical.
Signal (image) carrier	Visible light waves reflected from the surface.	Low energy, secondary electrons (SE) generated at the surface. High-energy, back-scattered electrons (BSE) reflected from the surface.	Transmitted electrons with variable energies.
Method of image capture and display	Optical images can be viewed through eyepieces or projected on a ground glass screen. Images may also be captured on photographic film or by a CCD; digital images from the CCD can be viewed in real time on a computer monitor or stored.	The electrons are collected and mapped as points of varying intensity to form an image. The image is recorded by striking a fluorescent screen, photographic film, or a CCD camera, which allows real-time viewing on a monitor or storage.	Output is seen as a diffraction pattern or a microstructural image. Both are recorded in the same manner as in the SEM.
Magnification	5× to 2000×	5× to 300,000× typical	50× to 1,500,000× typical
Resolution	<250 nm	3 nm typical (function of kV)	<0.2 nm typical (function of kV)
Information that can be obtained	Topographical characteristics of fracture surfaces. Microstructural characteristics of prepared sections.	Topographical characteristics of fracture surfaces. Microstructural characteristics of prepared sections.	Crystallographic orientation of phases (from diffraction pattern). Microstructural characteristics of sample.

TABLE 2.2 (*Continued*)

Typical Features of Optical and Electron Microscopes Used for Failure Analysis

	Instrument		
Basic Features	**Light Microscope**	**SEM**	**TEM**
		Detection of various phases in prepared sections, crystallographic orientation, and micro-strains (EBSD). Elemental composition of elements, generally from B (Z = 5) to U (Z = 92), with EDS or WDS.	Topographical characteristics of fracture surfaces via replicas.
Typical attachments for failure analysis	Micro-hardness tester. Atomic force microscope for very high-resolution surface topography characterization. Auto-montage stitching tool with auto-focus attachment.	Energy dispersive X-ray spectrometry (EDS). Detection limits of trace elements are up to about 0.1% wt (element specific). Wavelength dispersive X-ray spectrometer (WDS). Detection limits of trace elements are up to about 0.01% wt (element specific). Electron backscattered diffraction-analyzer (EBSD).	Various attachments are available, but are mainly used for research work.

indicate that the phase had promoted easy fracture. In the second instance, the nature of the corrosion products could indicate whether any particular species had assisted in the corrosion process, especially if localized cracking were present.

However, high magnification is not always necessary; in large samples where it is required to evaluate the overall characteristics of the component, for example, the solidification pattern of a casting, a low magnification would be required (Figure 2.4).

2.3.3 Mechanical Testing and Chemical Analysis

The engineering usage of materials is based upon specific properties of the material, which have been standardized in material specifications. Where the component is subjected to mechanical load, strength, ductility, and toughness would be the main considerations. In metallic materials, these properties would be dependent upon chemical composition and

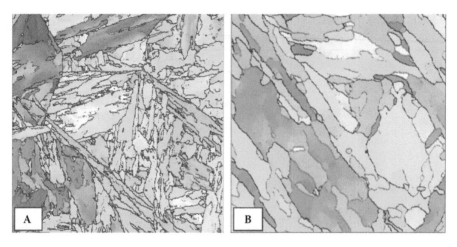

FIGURE 2.3
(See Colour Insert.) EBSD-IPF images of ASME A335, P92 steel. (With permission from Guat-Peng Ng et al. [1].) (A) Shows the normal fine-grained, tempered martensite phase required for creep service at elevated temperatures. (B) Shows that ferrite with inferior creep strength had formed after treatment just below the A_{C1} temperature.

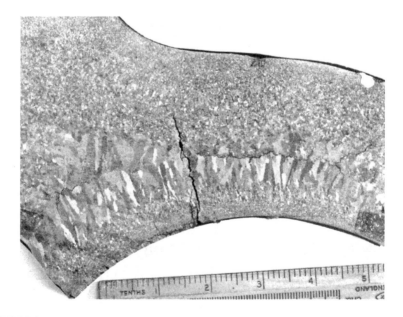

FIGURE 2.4
Macrograph of an austenitic manganese steel casting, with a coarse solidification structure and shrinkage cracks.

processing methods, whether cast, or wrought or heat treated, for example. A proper chemical composition allows suitable processing to be performed to obtain the required properties; the compositional requirement though can be judiciously relaxed in favor of mechanical properties, except in special cases where corrosion resistance is the main consideration. For example, a shafting steel may be specified as AISI 4140 with certain minimum properties, but if a higher grade steel with superior properties (AISI 4340) and a different chemical composition were to be supplied, there would be no adverse consequences to the performance of the shaft. The only consequence would be to the supplier's profit margin, as the AISI 4340 is a more expensive steel. But however, if an expensive AISI 316L stainless steel were to be specified for its superior corrosion resistance, usage of a cheaper AISI 304 should quickly attract unfavorable attention because the latter steel would have an inferior performance. All FIs should have a sound metallurgical education in order to be able to understand the basis of material usage more deeply.

2.3.3.1 Mechanical Testing

Numerous mechanical tests are available to characterize the mechanical properties of a given material. They are, in approximate order of complexity, hardness, bend, tension, compression, impact toughness, shear, fatigue, creep, and fracture mechanics (FM). Under normal circumstances, if there is a suspicion of inadequate mechanical properties, only the three primary properties used by design engineers, namely strength, ductility, and toughness would be tested (ductility and toughness are related but not equivalent). Strength and ductility are obtained in a simple, standardized, tensile test, which would give the yield stress, tensile strength, percentage elongation and reduction in area. For toughness, a V-notch Charpy impact test conducted over a range of temperatures is the most common. These tests are simple and can be made on representative samples in basic mechanical testing laboratories. In many standard specifications, hardness values are also specified, which can be performed in basic laboratories using standard Brinell, Vickers, or Rockwell machines. If there is a necessity for more advanced testing to define the parameters used in FM calculations, or to test at the extremes of temperatures or strain rate, specialized equipment and laboratories would be needed. These tests take longer to conduct and are much more expensive than the tensile, Charpy, or hardness tests.

The FI will very quickly discover that it is not always possible or convenient to perform tests on excised specimens. Firstly, there may not be sufficient unaffected material left for the test pieces to be excised and secondly, in large components made of hard steels or austenitic steels, for example, excising the material and machining the test pieces would be a fairly major task. In many cases, a competent FI with a good knowledge of metallurgy would be able to obtain a fast and sufficiently accurate characterization of the failed component by using portable hardness testers on the component, supported by microstructural examination via replicas. These tests would take a competent technician just a few hours to complete.

2.3.3.2 Chemical Analysis [2–3]

The aims of chemical analysis are to detect and quantify the elements present in the samples under investigation; this is performed on bulk material as well as on fine debris. Tests on fine debris can be performed by micro-analysis using SEM with energy dispersive X-ray spectrometry (EDS) or wavelength dispersive X-ray spectrometry (WDS) attachments (Table 2.2). In bulk analysis, the interest is mainly centred on whether the material has met its specified analysis, and whether its impurity levels had been high enough to have caused degradation in properties. The most commonly used instrument for metals is the spark-OES (optical emission spectrometer), which has advanced so much in the last few decades that modern instruments have detection limits down to a few ppm by weight and can detect elements from Li (Z = 3) to U (Z = 92). The test sample needs to be in solid form with a test area of about 10 mm diameter to cover the spark source; preparation just consists of hand grinding the surface to be tested, to a flat finish. Reproducibility is very good in a well-maintained and calibrated instrument and operation is easily performed by a trained technician. However, free graphite in cast irons cannot be accurately analyzed by spark-OES and a complementary method such as high-temperature combustion needs to be used. Other methods, which are less often used in the failure analysis of metals, include the wet methods, where the sample needs to be dissolved. This may be problematic in certain alloys that do not dissolve readily in chemical reagents. Wet methods include the classical 'wet analysis' where titrimetric quantification is performed, as well as atomic absorption spectroscopy (AAS) and inductively coupled plasma atomic emission spectroscopy (ICP-AES). These methods can have detection limits better than spark-OES but need more careful sample preparation and require a longer time for analysis, as well as a trained chemist to perform the task reliably.

2.3.3.3 Other Tests and Physical Modeling

Other tests may occasionally be necessary, such as advanced NDE to quantify internal flaws in a high-temperature vessel, or corrosion tests to determine whether a certain grade of stainless steel would suffer from stress corrosion cracking in seawater. It may sometimes be necessary to build small-scale models to test for effects of loading that are difficult to quantify by analytical calculations, for example, transverse wind loading on a long suspension bridge. The reader will have to explore other literature for a more detailed description of these topics, as they are beyond the scope of this book.

2.4 Stress Analysis and Computer Modeling

Stress analysis of components operating under normal environments is normally restricted to the mechanical modes of failure, namely, ductile fracture (DF), brittle fracture (BF), and fatigue failure (FF), some details of which have been given in Chapter 1. In cases where preliminary evidence suggests that the failure had been due to a lack of capacity of the component, but that its construction had followed design specifications, it would be necessary to establish whether it had been under-designed. This requires an analysis of the stresses in accordance with the prevailing design codes, to compare with the original stresses. In complex configurations, computer modeling may be necessary.

2.4.1 Ductile Fracture

Normal design methodology assumes that the component would fail in a ductile mode, meaning that its yield stress would be exceeded and that it would exhibit macro-plasticity before failure; this would mean that the component has reached its maximum capacity before failing. The design would then aim to allow the presence of certain combinations of stresses that would be lower than the capacity of the component by a certain amount, a so-called factor of safety. Many such components are governed by codes of practice (CP) approved by the industry and regulatory bodies, which must be followed. Examples include structural components such as bridges and buildings, and mechanical components such as pressure vessels and boilers. Certain other components are governed by in-house practices and performance requirements; examples include vehicles and machinery. Where a CP exists, re-analysis of the design is quite straightforward but in its absence, analysis needs to start from scratch. For simple configurations, manual calculations can be made, but for complex ones manual calculations would not give answers fast enough or accurately enough to be useful. Fortunately, there are now finite-element-based computer programs available to help with even the most difficult analysis, in addition to specialists for hire to run the programs. However, all the input comes from the FI.

2.4.2 Brittle Fracture

On the other hand, there are failures that occur at stresses lower than the designed capacity of the component, by BF, without prior signs of distress; these conditions are described as fully elastic or plane strain. The methodology of BF control is different from that used for DF control because there is no single material property that can be used to characterize BF, unlike the use of the yield strength to characterize DF. And even when tensile tests have shown that a material has a large elongation at room temperature, the material may still fail by BF in the field. Generally, BF behavior is dependent upon the thickness of the component, the temperature of operation, the presence of stress concentrations and residual stresses, the toughness and microstructure of the material, and the rate of application of stress. There is no single methodology that can encompass all these variables, and techniques of BF control have evolved into two forms, a transition temperature approach and a fracture mechanics approach.

The transition temperature approach is normally applied to metals or alloys with a body centered cubic (BCC) unit cell (such as ferritic/pearlitic steels). Such materials fail by ductile mechanisms of MVC (microvoid coalescence) and shear at elevated temperatures but by brittle mechanisms of cleavage at low temperatures. The transition temperature depends upon factors such as grain size, chemical composition, impurity content, mechanical strength, and thickness of section. The BF control method is then used to establish a reproducible characteristic temperature that can be obtained with a suitable test. Many test methods have been proposed in the past but the Charpy V-notch impact test is the most widely recognized and the most commonly used presently. Many CPs specify that the steel used (structural grade) should have a certain minimum Charpy impact energy (for example, 27J) when tested at the lowest expected temperature of operation. This requirement has served very well to reduce the frequency of BF. So, if such a steel were to fail by BF, samples from the failed component could be subjected to Charpy testing to see if it was suitable for use.

The transition temperature methodology is not applicable to metals and alloys with a non-BCC unit cell or those with high-strength and low-toughness properties, which include widely used structural materials such as Al and Ti alloys, austenitic stainless steels, and high-strength martensitic steels. For such materials, fracture mechanics (FM) methodology is currently used and the LEFM equation (Equation 1.1) in Chapter 1 would be applicable, where the fracture stress σ_c can be calculated if the critical crack size a_c, the plane strain fracture toughness K_{IC}, and the Y function are known. The Y functions for various configurations may be taken from handbooks [4], though there are now computer programs available to calculate them [5–6]. However, Equation (1.1) in Chapter 1 is only valid when the thickness (t) and other characterizing dimensions are sufficiently high to induce fully elastic, plane strain conditions. The thickness condition for plane strain has been determined to be [7],

$$t \geq 2.5 \left\{ \frac{K_{IC}}{S_y} \right\}^2 \tag{2.1}$$

where Sy is the yield stress of the material.

If the thickness condition cannot be satisfied, elastic-plastic conditions would exist at failure, and the capacity of the component would lie between the plane strain capacity (minimum value) and the condition for full plastic yield (maximum value under ductile conditions). Then, methodologies involving, for example, the *J integral* or the *R-curve* would need to be applied.

2.4.3 Fatigue Failure

Fatigue failure can occur at low stresses to give HCF (high-cycle fatigue) or at very high stresses to give LCF (low-cycle fatigue); the stresses must, however, contain a variable component. It is the most common mode of mechanical failure worldwide, to such an extent that many insurance companies specifically exclude such failure from coverage. The quantitative analysis of fatigue would need to consider a host of parameters such as the nature of the applied forces, the effects of stress concentrations and residual stresses, the fatigue strength of the component and its susceptibility to size effects, surface roughness, and operating environment. Advanced techniques would be needed to analyze crack growth rate and cumulative damage and understandably, it is the most difficult of the three mechanical modes to treat quantitatively. Analysis methodology includes the commonly called stress-life approach [8], the strain-life approach [9], and the fracture mechanics [7] approach.

2.4.3.1 Stress-Life Approach

This approach is the oldest and most easily used and has been exhaustively treated in many excellent texts [10]. It is quite accurate for HCF but not for LCF, where large plastic strains are involved. The approach essentially seeks to obtain a stress level that is less than the fatigue strength of the component. The fatigue strength is derived from the S-N curve of the component or a component-like test member. For steel, the strength could be at the finite region, approximately between about 10^3 and 10^7 cycles for failure occurrence, or at the infinite region beyond 10^8 cycles, where no failure occurs. If the fatigue strength of the component were not available, which the FI would find to be usually the case, approximations may be made by applying reduction factors to the endurance limit for the material, which is measured on small, standardized test samples. Where the endurance limit is not available, approximations can be based on the tensile strength of the material. These approximations are of course subject to some error, but at least a first approximation can be made. In cases where there is a mean stress (σ_m) and a variable stress (σ_a) component, a suitable failure criteria needs to be used. These range from the conservative, Soderberg straight line criterion to the Gerber parabolic line and to the ASME-elliptic line [10]. The Soderberg and Gerber lines were derived very early in the history of fatigue analysis and the ASME line is the most recent. The Soderberg and ASME-elliptic lines are given as examples below:

Soderberg:
$$\frac{\sigma_a}{S_e} + \frac{\sigma_m}{S_y} = \frac{1}{n} \tag{2.2}$$

ASME-elliptic: $$\left(\frac{n\sigma_a}{S_e}\right)^2 + \left(\frac{n\sigma_m}{S_y}\right)^2 = \frac{1}{n}$$ (2.3)

where,

S_e is the fully corrected endurance limit (fatigue strength) at the critical
        location of the component,
S_y is the yield stress of the material, and
n is the design factor or factor of safety.

Equations (2.2) and (2.3) describe loci when plotted on rectangular coordinates, which separate a safe region from an unsafe one.

2.4.3.2 Strain-Life Approach

This is considered to represent the fatigue process more accurately than the stress-life approach, especially in the LCF regime. In this case, the local stresses (σ) and strains (ε) at notches are estimated and used as the basis for life estimation. The S-N curve would be for strain life and can be represented by the Manson–Coffin relationship [11–12] as,

$$\varepsilon_a = \frac{\sigma'_f}{E}\left(2N_f\right)^b + \varepsilon'_f\left(2N_f\right)^c$$ (2.4)

where,

ε_a is the strain amplitude,
The first term on the right-hand side (RHS) is the elastic strain component,
The second term on the RHS is the plastic strain component,
N_f is the cycles to failure for completely reversed loading (also often
        regarded as the cycles to crack initiation),
σ'_f, ε'_f, b, and c are material fatigue constants, and
E is the elastic modulus of the material.

In practical usage, it is difficult to determine with accuracy the total strain at the bottom of a notch, compounded with the fact that there is, at present, little published literature on such values. Various approximations are however available.

2.4.3.3 Fracture Mechanics Approach

The stress-life and strain-life approaches attempt to define a safe life below which failure would not be expected to occur and they do not treat practical situations where growing Stage II fatigue cracks are present. The FM approach does this well and provides a basis to make a decision on when to take such cracked components out of service without sacrificing safety. The rate of crack growth then needs to be known, and it can be represented by the Paris equation [13], which takes the form,

$$\frac{da}{dN} = C(\Delta K)^m \qquad (2.5)$$

where

a is crack length, N is cycle, $\dfrac{da}{dN}$ is crack growth per cycle,

K is stress intensity factor and $\Delta K = K_{max} - K_{min}$, the stress intensity range.

Equation (2.5) is however too simple to account for the effects of the stress ratio R (maximum load divided by minimum load), and is inaccurate in very low-and very high-K regimes. More accurate (but more difficult to use) equations have been formulated, which take these factors into account, for example [14],

$$\log \frac{da}{dN} = C_1 \sinh \left\{ C_2 \log (\Delta K) + C_3 \right\} + C_4 \qquad (2.6)$$

where,

C_1 is a material constant, and C_2, C_3, and C_4 are functions of the stress ratio (R), frequency, and temperature.

The use of Equations (2.5) and (2.6) is to estimate the number of cycles that are needed before the crack length becomes large enough to precipitate total failure. This knowledge provides a sound basis for allowing the component to operate until the crack length approaches critical dimensions, which needs to be estimated from the approaches mentioned in Sections 2.4.1 and 2.4.2. In the FM method, periodic NDE must be carried out to validate the calculated crack growth rate and operation must be closely monitored to ensure that no sudden excursions of load occur beyond those used in the critical crack estimations. Experimental plots of $\log \dfrac{da}{dN}$ with $\log(\Delta K)$ will show that a threshold value, ΔK_{th}, would be present, where crack growth tends to be zero. It would seem that if it were possible to operate at regimes below ΔK_{th}, there would be no crack growth, but this is not a viable proposition, as the applied stresses need to be so low that it would be impractical.

2.4.3.4 Cumulative Damage

In all three approaches discussed, attention would need to be given to situations concerning finite life regimes where variable amplitudes occur at different periods. Palmgren–Miner [15–16] proposed a cycle-ratio summation rule to calculate the cumulative damage, which is still widely used today:

$$\sum \frac{n_i}{N_i} = c \qquad (2.7)$$

where n_i is the number of cycles at stress σ_i and N_i is the number of cycles to failure at this stress. The parameter c has been found experimentally to range between 0.7 and 2.2, with an average value near unity. Based on these, a linear damage rule to estimate the accumulated damage D, can be written as,

$$D = \sum \frac{n_i}{N_i} \qquad (2.8)$$

Failure is predicted when $D = c = 1$, though to be conservative, lower values than unity may be used.

2.4.3.5 Preliminary Qualitative Assessment

The FI should not hastily embark on complex stress analysis without first applying some common engineering sense. For example, in a single component made of a tough, ferritic/pearlitic steel, if only a small (with respect to total stress area) Stage II fatigue crack had precipitated a ductile fracture, it would mean that the component had insufficient strength capacity and a stress check would then be indicated. Conversely, if the Stage II crack was large, and there was no load transfer mechanism to other connected members, it would mean that the component had a high-strength capacity. Then, the FI should look for high stress concentrations that had promoted easy crack initiation and propagation, and a stress check would be superfluous. In the case of high-strength martensitic steels, a small crack resulting in brittle fracture with very small shear lips would point to low toughness (K_{IC}) of the material.

2.5 Report Writing

2.5.1 Objectives of a Failure Analysis Report

A failure analysis report is a technical document that is aimed at conveying technical information to the client and also to serve as a record for future reference, if needed. In this book, we are only concerned with failure or forensic analysis reports, but the writing style is applicable to all technical reports. All such reports may have legal implications and the authors of such reports must be fully aware of their legal responsibilities and obligations. The reports are usually for specific clients and the report structure will vary according to circumstances. Reports on failure analysis are difficult to write well and can have many different but equally acceptable formats. Failure analysis reports are normally used as a basis to prevent future failure occurrences, for insurance liability claims as well as to identify the parties at fault, for litigation proceedings.

2.5.2 General Report Structure

2.5.2.1 Overview

The FI needs to tell the reader what they intend to do, how they did it, and the results that were obtained. The general report structure may include some or all of the following parts and certain sections may be combined or expanded, depending upon the nature of the problem and the needs of the client involved.

1. Executive summary
2. List of contents
3. Introduction and background information
4. Work scope and methodology
5. Observations and test results
6. Discussions
7. Conclusions
8. Recommendations for further action
9. Figures and tables, references, appendices, and attachments

2.5.2.2 Executive Summary

The executive summary is often considered the most important section of a report. It should highlight the strengths of your overall report and therefore be the last section you write. However, it usually appears first in your report document. In a lengthy failure analysis report, the executive summary is one which contains the context of the investigation, the purpose of the report, the major findings, the conclusions, and the main recommendations. A busy manager/client and one who probably has very little engineering knowledge will expect to understand the gist of the problem by just reading the executive summary.

2.5.2.3 List of Contents

The contents of the report should be spelled out at the beginning of the document for easy reference and retrieval of information. In a long report, this gives more detailed information than a formal work scope.

2.5.2.4 Introduction and Background Information

This section provides sufficient background information of the problem at hand, summarizes what is done and why, and the writer's role in solving the problem. The information generally includes the following:

1. Function, specifications, design and construction, and age.
2. Immediate events leading to the discovery of the failure incident.
3. Operational and maintenance practices.
4. Recent abnormalities in the machinery and/or human behavior.

The sources of information would include, but not be limited to the following:

5. Written reports on the problem from various sources (in-house reports, police, insurance, and fire department).
6. Verbal information and signed witness interrogation accounts.
7. Tender documents, equipment specifications, operation and maintenance manuals, certificates of commissioning, etc.
8. Operational and maintenance records, logbooks, chart recordings, and computer-monitored parameters.
9. Literature surveys from published sources such as books, research papers, journals, and the Internet (though this source must be used with care).

2.5.2.5 Work Scope and Methodology

The *Work Scope* should state the extent of the work to be undertaken, and for contractual reasons it should not be less than what has been mutually agreed upon and should not contradict the client's instructions. If, for reasons of professional ethics, the FI does not agree with the instructions given, they should reject the assignment. The *Methodology* should state how the FI intends to carry out the work. For work where the FI's test results will be used as judgment criteria for acceptance or rejection of equipment or of work done by other parties, the methodology should be rigorous and should conform to mutually agreed protocol.

2.5.2.6 *Observations and Presentation of Test Results*

This section of the report reveals what has been observed, discovered, and analyzed. The subject matter should include:

1. Evidence from site work from activities such as visual examination, thickness gauging, hardness testing, replica extraction metallography, NDT, PMI.
2. Types of samples acquired for laboratory testing.
3. Evidence from laboratory examination and testing such as visual and macro-examination, optical and electron microscopy (SEM and EDS), mechanical and material testing, and special tests.
4. Presentation and analysis of results of test data using tables, graphs, or charts. If lengthy, they should be placed in appendices, with only a summary in the main text. Analysis of numerical data should generally follow normal statistical methodology, with the range, mean, and standard deviation included. Analysis of non-numerical observations is more subjective but nevertheless should be made as accurately and as objectively as possible. Pure observations should be clearly differentiated from your opinions.

2.5.2.7 *Discussions*

The following areas should be interpreted and discussed clearly:

1. The nature of the failure (factual observations)
2. Mechanisms of failure (factual observations)
3. Possible proximate causes (your deductions)
4. Possible root causes (your deductions)

In the discussion of the results, details should be cited from the findings, sources, and the theory as evidence of specifics in the claims. Use the present tense to make generalizations. A good test of the accuracy of your deductions would be to construct the most likely scenario of occurrence and to subject critical stages to 'what if' sub-scenarios to see if the challenges can be met. A fault tree containing all possible scenarios can also be constructed but this may incur an unnecessary waste of time and effort, as it would require an analysis of all the possible failure modes of the component.

2.5.2.8 *Conclusions*

The final conclusion should sum up the main points of the investigation. It should clearly relate to the objectives of the report. There should be no surprises, that is, do not include new information here.

2.5.2.9 Recommendations

Appropriate recommendations may be made to point the way forward to prevent further recurrence of the problems. They must be logically derived from the results of the investigation.

2.5.3 Writing Style and Content

2.5.3.1 Language

A report aims to inform, as clearly and succinctly as possible. It should be easy to read, and professional in its presentation. Specifically, it is meant to convey information efficiently and unambiguously. A technical report is not a newspaper or magazine article or some insurance clause and should not be written in the same manner. Flowery and complicated language should be avoided where the use of short, simple, and direct sentences will serve the purpose better. This is particularly important where the report is written in English and is also read by non-native English speakers. If any dispute were to arise it should preferably be in the interpretation of the results rather than in the interpretation of the meaning of the words used. If insurance clauses, or even the Constitution of some countries, were to follow these guidelines, there would be less need to bring disputes to a court for a decision to be made on what the wordings exactly meant. Some elegance in language would of course give an impression of scholarship and elicit more respect from readers than a crudely crafted document. There is no single best way to write a report, but all good reports possess clarity, precision of language, continuity of subject matter, and objectivity. The limitations of your report should be clearly stated.

1. *Clarity* is achieved by organizing the information under appropriate topics with sub-headings and proper paragraphs. Take note that clarity must be from the reader's perspective and not the writer's, as the writer is familiar with the subject but the reader may not be.

2. *Precision of language* requires a good command of the language, which can be achieved by conscious practice. Always think of what is needed to be said and to say it exactly using short simple sentences wherever possible. Write to express not impress.

3. *Continuity of subject matter* is achieved by linking up sections, sub-sections, and paragraphs coherently and intelligently, so that the narrative unfolds naturally.

4. *Objectivity* is essential in the report in order to maintain integrity and to be able to withstand challenges in court. Deliberate deviations from the truth will place the parties involved in jeopardy of heavy penalties and damages.

5. A *passive voice* is used in conventional reports because the event or action is more important than the doer. However, in cases where

personal opinion or verification is crucial to the event the active voice should be used. Currently, the plural form 'we' is used instead of 'I' and the plural 'they' for 'he/she'.

6. *A rigorous but simple and easy to read report* allows a non-specialist to fully understand the contents, but is rigorous enough for a specialist to know that the investigator is competent to deal with the problem.

7. *Abbreviations (e.g., Cu), acronyms (e.g., ASME), technical terminology, and scientific symbols* are inevitably used in technical reports, which would be a source of confusion and irritation to any reader who is not a specialist in the particular topic. These terms should be explained early in the report before any of them are used.

2.5.3.2 Documentation

Report writers should be painstaking in preparing the documentation because their reports depend substantially on the evidence that they have adduced from physical observations. Photographs, sketches, notes, or a voice recorder should be included as evidence. In failure analysis, a good picture is truly worth a thousand words (see the numerous illustrations provided in this book). Today, moderately priced, high-quality digital cameras are available, which can take large numbers of high-quality photographs. There is no reason whatsoever for not providing good illustrations in failure analysis reports, except for a lack of competence. As a precaution, legal issues may however surface over the use of digital photographs as digital files can be easily altered and digital images are often edited to lighten shadows, darken excessive highlights, and improve clarity. To maintain proof that the digital images have not been falsely altered to show features not originally present, digital files should be stored in duplicate on separate hard disks and in their original form with their Exif metadata intact. The relevant specialists may ask that these images be examined to prove their originality.

2.5.3.3 Disclosed Information

The report should only divulge, disclose, and impart information that is relevant to the case. Added but unnecessary information will not enhance the report but instead may cause it to lose credibility if the information happens to be erroneous. Information may be communicated by means of text, tables, graphs, and photographic illustrations, or by whichever method is the best to state the case and for the reader to understand fully. It is not good practice to insert numerous supporting figures or illustrations within the text as these can disturb the flow of the narrative and their location often needs to be readjusted when there are substantial changes to the text. However, if they are grouped together at the end of the report, they can be easily located and referred to.

2.5.3.4 Avoidable Mistakes

1. Factual evidence (objective observations, measurements, and test results) should be differentiated from personal opinions (subjective analysis). State the factual evidence first before using them to form personal opinions and ensure that the reader can easily distinguish between the two sides.

2. Ambiguous or vague statements will allow the reader to draw conclusions different from what is intended. This shows a lack of clarity in thought and should be avoided.

3. If the evidence is not certain or a satisfactory conclusion cannot be made, it should be clearly stated rather than commit an error or falsehood.

4. The use of empty words, phrases, and sentences should be avoided. For example, stating that, 'My results are inaccurate because of some problems with my testing machine' will not satisfy a discerning client, or the court.

5. It is bad practice to dump a mass of raw data in the appendix and expect the reader to read through and analyze it. It is the writer's responsibility to perform the analysis and to present the findings. The data is there to show that it actually exists and to allow a check to be made on its accuracy and on the accuracy of the analysis, if so needed.

6. The use of acronyms/abbreviations makes typing easier and reduces reader fatigue. For example, 'tertiary superheater tube No. 5' can be reduced to 'TST 5'. If the full phrase has to be repeated 20 times in a report, much time will be wasted.

7. Careless mistakes should not be made on simple matters that will make the reader lose confidence in the analysis of the important issues.

2.5.3.5 Assurance of Report Quality

All reports, technical or otherwise, should ideally be perfect in every way but this state is seldom achieved. However, there must be some assurance of absence of serious mistakes that would negate the findings of the report. This can be achieved by observing a few precautions; firstly, the FI/writer of technical reports should be competent in the topic and possess the necessary academic qualifications and professional experience to undertake the assignment. In works of an engineering nature, these requirements are mandated by law in most countries. Secondly, there must be access to good and speedy laboratory facilities, whether in-house or from third parties. Thirdly, there must be access to good library facilities and to experienced personnel for discussions and exchange of opinions. And lastly, the finished draft should be carefully reviewed for errors. It is best that the review be performed by another competent person, but if this is not possible then the author is well advised to leave the manuscript for a few days before re-checking, as errors tend to become undetectable after one has spent long hours looking at and amending the same report.

2.5.3.6 A Final Note of Advice

The author of this chapter has more than four decades of experience as a university lecturer, professional engineer, and failure investigator. He has found that failure analysis reports are the most challenging of all technical reports to write well. The main difficulty most times lies in the process of linking all the observations and results from the different stages of the investigation so that the report unfolds naturally. He does this by first preparing the photographic illustrations so that they form a skeleton of the report, to be fleshed out by the descriptive text and supported by attachments and appendices. The photographic evidence should show as far as possible the nature of the failure and the possible mechanisms of failure. For this reason, the FI or someone equally competent should be the one to take photographs of the failure scene and the mode of failure. This is because the various marks caused during failure may not be fully revealed if the shooting angle, point of focus, or lighting, are not all correct.

Problems and Answers

Problem 2.1

Visual examination and metallurgical examination are both important steps in obtaining physical evidence. Explain why visual examination needs to be performed prior to metallurgical examination. List four objectives of visual examination.

Answer 2.1

A detailed visual examination in the laboratory should first be performed to reinforce observations made at a site, to determine whether further site visits are needed to better understand the site conditions, or if more samples are required for testing. This would avoid samples from being damaged unnecessarily. It would be unwise to proceed beyond the visual examination stage and to start cutting up indiscriminately to obtain samples before the proximate causes have been identified to a fair degree of certainty.

Visual examination objectives:

1. The manner of deformation, whether caused before or during failure.
2. The presence of wastage and degradation, whether from normal wear and tear, abuse in service, corrosion, or erosion.
3. The fracture characteristics, whether caused by ductile, brittle, fatigue, or other mechanisms.
4. Locations of crack initiation where fracture/failure had been preceded by cracking.

Problem 2.2

Brittle fractures and ductile fractures are two of the best-known failure modes. Explain the main macroscopic differences between them.

Answer 2.2

Brittle fracture involves relatively little plastic deformation and normally occurs suddenly without any warning signs. Brittle fracture paths tend to be perpendicular to the principal tensile stress. The brittle crack is unstable and propagates rapidly without an increase in applied stress. The fracture surface has characteristic chevron marks and in ferritic/pearlitic steels is bright and shiny in appearance.

On the other hand, ductile fracture involves prior yielding and plastic deformation on a large scale. The progression to final failure may take some time to occur, which allows safety measures to be taken. Crack paths could be parallel or inclined at 45° to the direction of maximum shear stress. If parallel, the fracture surface would normally have a smooth, silky appearance. If inclined, the appearance would normally be dull and fibrous.

Problem 2.3

You are in an isolated site with only a hardness testing machine, a vice, and some rudimentary tools. You have a length of steel of about 20 × 20 mm section that must be used to urgently replace a broken member. What tests can you perform to ensure that the steel will be adequate?

Answer 2.3

Firstly, carry out hardness tests on the steel and convert the hardness value to tensile strength. (What formula would you use?) If the strength were adequate, the next step would be to gauge its ductility. For this, you need to cut out a suitable piece and bend it in the vice using a hammer. If it can bend to 90° without cracking, the ductility would be sufficient. Note that an experienced

FI with a set of suitably graded files of different hardnesses would be able to gauge the hardness fairly accurately in the absence of a hardness machine, but of course, accurate portable hardness testers are now easily available.

Problem 2.4

Which industries use the API and ASME codes?

Answer 2.4

API codes are widely used in the oil and gas industries.

The ASME Boiler and Pressure Vessel codes are widely used in the power generating industries as well as in industries using pressurised components.

Problem 2.5

You need to go to a site, which has good lighting conditions to take photographs of a failure where the components are large in size and space for movement is very restricted. You are only able to carry one camera; what sort of camera would you take and why?

Answer 2.5

A small sensor digital camera (say, 1/2.3″ sensor size) with a minimum focal length of 25 mm and a minimum zoom ratio of 15× would be quite suitable. The small sensor will have a large depth of focus; the 25 mm focal length will allow fairly wide-angle shots; and the 15× zoom will allow close-ups to be taken at a distance. Large sensor cameras (micro four-thirds and above) would give better resolution but their depth of field and zoom ratio would not be adequate.

Problem 2.6

(a) Why is it that large steel components made of high-strength martensitic steels sometimes fail by brittle fracture, but yet have high ductility in a tensile test specimen? (b) What tests would ensure safety from brittle fracture? For API pipelines, what method is recommended?

Answer 2.6

(a) The large component is under plane strain conditions, which favor brittle fracture, whereas the small tensile test specimen is under plane stress conditions where ductile failure prevails, so this test cannot replicate performance under actual conditions.

(b) A fracture mechanics test to obtain the K_{IC} value of the steel would be necessary. For API pipelines, a transition temperature approach using the Drop Weight Tear Test is recommended (API RP 5LR or ASTM E436). This approach is not suitable for high-strength martensitic steels, which do not exhibit pronounced transition temperature behavior.

Problem 2.7

How would you use Equation (2.5) to estimate the end of life of a component with a growing fatigue crack?

Answer 2.7

(a) Obtain values of the required material properties.

(b) Measure the crack profile and depth as accurately as possible.

(c) Determine the component and load configuration and calculate the appropriate K value.

(d) Carry out numerical integration for the rate of crack growth.

(e) Measure the crack length at periodic intervals to verify that the crack growth rate has been accurately predicted.

(f) Determine the critical crack length for plane strain or plane stress conditions (whichever is applicable) and the number of cycles required to reach this condition.

(g) In practical usage, the component needs to be taken out of service before the critical crack length has been reached and an appropriate reserve margin is applied.

References

General References

ASM Handbook, Vol. 8, *Mechanical Testing and Evaluation*, Volume coordinators, H. Kuhn and Dana Medlin, ASM International, Ohio, 2000.

ASM Handbook, Vol. 9, *Metallography and Microstructures*, Ed., G.F. Vander Voort, ASM International, Ohio, 2004.

ASM Handbook, Vol. 12, *Fractography*, R.L. Stedfeld et al., ASM International, Ohio, 1987.

ASM Handbook, Vol. 13A, *Corrosion: Fundamentals, Testing, and Protection*, Ed., S.D. Cramer and B.S. Covino, Jr., ASM International, Ohio, 2003.

Specific References

1. Ng, Guat-Peng and Jung-Chel Chang, Study of the Creep Rupture Strength for P91 & P92 Steels after Short-Term Overheating above A_{C1} Temperature, The 5th Symposium on Heat Resistant Steels and Alloys for High-Efficiency USC/A-USC Power Plants, Seoul, Korea, May 2013.

2. Aliya, D., *Chemical Analysis of Metals in Failure Analysis*, in ASM Handbook, Vol. 11, *Failure Analysis and Prevention*, Ed., W.T. Becker and R.J. Shipley, ASM International, Ohio, 2002.

3. ASM Handbook, Vol. 10, *Materials Characterization*, Volume coordinator, R.E. Whan, ASM International, Ohio, 1986.

4. Tada, H., P.C. Paris, and G.R. Irwin, *The Stress Analysis of Cracks Handbook*, 2nd ed., St. Louis, MO: Paris Productions Inc., 1985.

5. AFGROW, U.S. Air Force Research Laboratory, http://fibec.flight.wpafb.af.mil/fibec/afgrow.html.

6. NASGRO Fracture Mechanics and Fatigue Crack Growth Analysis Software, Southwest Research Institute, http://www.nasgro.swri.org/.

7. Antolovich, S.D and B.F. Antolovich, *An Introduction to Fracture Mechanics*, in ASM Handbook, Vol. 19, *Fatigue and Fracture*, S.R. Lampman et al., ASM International, Ohio, 1996.

8. Kaplan, M.P. and T.A. Wolff, *Fatigue-Life Assessment*, in ASM Handbook, Vol. 11, *Failure Analysis and Prevention*, Ed., W.T. Becker and R.J. Shipley, ASM International, Ohio, 2002.

9. Mitchell, M.R., *Fundamentals of Modern Fatigue Analysis for Design*, in ASM Handbook, Vol. 19, *Fatigue and Fracture*, S.R. Lampman et al., ASM International, Ohio, 1996.

10. Budinas, R.G. and J. Keith Nisbett, *Shigley's Mechanical Engineering Design*, 8th ed., New York: McGraw-Hill, 2008.

11. Dowling, N.E., *Estimating Fatigue Life*, in ASM Handbook, Vol. 19, *Fatigue and Fracture*, S.R. Lampman et al., ASM International, Ohio, 1996.

12. Manson, S., *Proc. Heat Transfer Symp.*, Univ. Michigan, Eng. Res. Inst., 1953, 9–75.

13. Paris, P.C. and F. Erdogan, *J. Basic Eng. (Trans. ASME)*, Series D, Vol. 85, 1963, 528–534.

14. Wallace, R.M., C.G. Annis, Jr., and D. Sims, *Report AFML-TR-76-176, Part II*, 1976.

15. Palmgren, A., Die Lebensdauer von Kugellagern, *ZVDI*, Vol. 68, 1924, 339–341.

16. Miner, M.A., Cumulative Damage in Fatigue, *J. Appl. Mech.*, Vol. 12, *Trans. ASME*, Vol. 67, 1945, a159–a164, 159–164.

3

Transportation Infrastructure

3.1 Introduction

The primary concern of infrastructures, especially the foundation that supports or maintains mass transportation facilities, is safety. This is the key focus of all large-scale transportation infrastructure projects, that is, roads, highways, and trains including the mass rapid transit (MRT), light rail transit (LRT), and bus rail transit (BRT). Failure in road and rail networks has serious ramifications especially when human lives are lost. One such disaster occurred at the Penang Ferry Terminal in Butterworth, Malaysia on 31 July 1988, when the supporting steel structures in the bridge collapsed because of overloading, killing 32 people and injuring 1634. Common modes of structural failure in transportation infrastructure include cracks, deformation, corrosion, ground settlement, fatigue, and fracture. These catastrophes are caused by factors such as improper or inadequate selection of materials, overloading, faulty design, and natural calamities.

In this chapter, three case studies of transportation infrastructure failures are presented. The first case involves a train derailment in Malaysia in 1987. Rail track specimens retrieved from the site revealed fractures of various natures. Metallographic examination was conducted to inspect the rail specimens for variations in grain size and pearlitic/ferritic composition of both the weld and parent metals. Mechanical testing was also performed using the Charpy V-notch method. Stress calculations were obtained to verify the results. Key outcomes of this rail failure investigation indicated undetected weld defects as the cause of the fracture. The second case concerns the failure of the ro-ro ramp at a wharf in a port in Malaysia. Results of the macro and metallographic examinations revealed that the failure was precipitated by dislodgement of the threaded end of the support bolt at the port piston, from the female nut. Because of corrosion, especially at the top end, only 30% of the female threads in the nut were effective, thus reducing its capacity to 81% of the maximum design load of 2551 kN. In the third case, a girder launcher collapsed during construction of the Kuala Lumpur's light rail transit (LRT) system. The lower hinge connection of the left girder, which had grossly deficient weld metal broke from excess stress load.

3.2 Case Study 1: Welding Defects in a Rail Track

3.2.1 Background

In the late 1980s, a train derailed in the state of Johor, Malaysia. The train was traveling from south to north; the left and right rails are defined when looking in the direction of travel. Samples were taken from the broken left rail track for examination (Figure 3.1). There was a fracture at the thermit welded joint (TWJ) at location B. There was another fracture at location A on the parent metal. Figure 3.2 shows a close-up view of the fracture surface

FIGURE 3.1
Samples are from the left rail track; the north direction is to the left of the figure and south is to the right. Arrow C points to the north fracture, and B points to the south fracture at the TWJ. Arrow A points to a deformed, broken-off piece between B and C.

FIGURE 3.2
A close-up view of the fracture surface at the TWJ (Figure 3.1B), on the south length of track, containing a large, darkish, pre-existing crack.

FIGURE 3.3
The south length of track, containing the fracture surface at the TWJ, is shown magnified in Figure 3.2. The track length south of the fracture surface was undeformed over 3 m or so.

FIGURE 3.4
The north length of track, beyond length A, shows fracture C with ductile characteristics caused by an overstress failure.

on the TWJ, on the sample south of B; it contained large areas of darkish pre-existing defects. Figure 3.3 shows that the rail, which was south of the TWJ, that is about 3 m in length, was in an undeformed condition.

The fracture surface at point C, on the north length of track, contained characteristics typical of a ductile, overstress failure (Figure 3.4). It then became clear that the fracture at the TWJ had occurred first; and then two points remained that needed to be answered:

    a. What manner of stress had caused the TWJ fracture to occur?
    b. Was the fracture the cause or the result of the derailment?

3.2.2 Method of Investigation

A specimen of the fractured left rail track was removed from the site and subjected to macro, metallographic, and mechanical examinations. The macro-examination looked for physical attributes of the damage, conditions (including defects) of the weld and parent metals, as well as growth of fatigue and brittle cracks especially in the vicinity of the TWJ. The metallographic examination studied the microstructure of the weld and parent metals in the rail tracks while the mechanical testing confirmed whether the weld and parent metal properties fell within the expected range of BS 11 steel [1].

3.2.3 Results

3.2.3.1 Macro-Examination

A part of the rail track in the TWJ containing the fracture shown in Figure 3.2 was examined. Left unprotected and exposed to the elements, the fracture surface was covered with a thick layer of rust (Figure 3.5), so loss of vital information concerning the fracture mechanisms would be expected. Figure 3.6 shows the fracture surface after it was partially cleaned. Figure 3.7 indicates the seven regions visible on the fracture surface. They are:

1. *Region A*: Weld material did not fully penetrate the parent metal.
2. *Region B*: Joint between the weld and parent metals contained a large number of blowholes.
3. *Region C*: Fatigue cracks propagated from the initial defects in region B.
4. *Region D*: Weld defect.
5. *Region E*: Possibly caused by a recent fatigue crack propagation.
6. *Region F*: Brittle fracture at the foot of the rail caused by cleavage mechanisms and direction of growth (Figure 3.7).
7. *Region G*: Brittle fracture at the web and head of the rail also caused by cleavage mechanisms and direction of growth (Figure 3.7).

The manner of deformation at the top edge of the head is illustrated in Figure 3.8. This suggests that the final fracture had been caused by vertical wheel loads acting downwards, suggesting that the last part to fail was G.

3.2.3.2 Metallographic Examination

Various specimens were extracted at different points across the fracture planes at the foot, web, and head of the rail tracks. The parent metal revealed a coarse-grained pearlitic/ferritic steel (Figure 3.9) with a hardness value of approximately 238 HB, conforming to BS 11. The weld metal was also essentially pearlitic/ferritic, having different grain sizes and different proportions

FIGURE 3.5
Fractured surface covered with rust (the same part as in Figures 3.2 and 3.3).

FIGURE 3.6
Partially cleaned TWJ (the same part as in Figure 3.5).

of pearlite and ferrite, with hardnesses ranging from 280 to 237 HB at various locations (Figure 3.10). These values showed fairly large variations but in and of themselves would not have had a significant impact on the rail performance. Failure in region C (Figure 3.7) was due to fatigue crack growth while failure in regions F and G were the result of brittle fracture. The characteristics contained in region E could not be ascertained, but this

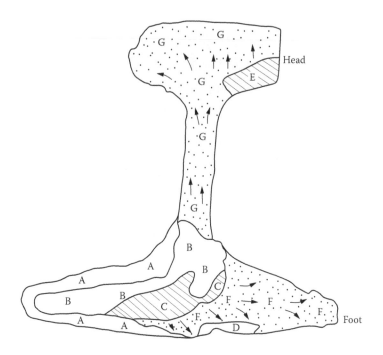

FIGURE 3.7
Regions A to G on the fracture surface with arrows indicating the direction of the final (brittle) fracture.

FIGURE 3.8
Local deformation at the top edge of the rail head, at the fracture shown in Figure 3.5.

was of little consequence as that part occurred at the end of the fracture process. A micro-section of the head of the rail revealed the presence of a fatigue crack of approximately 10 mm long and filled with oxides. The crack was located approximately 18 mm from the fracture plane, showing that it had started a long time ago (Figure 3.11).

FIGURE 3.9
The parent metal was a coarse-grained pearlitic/ferritic steel.

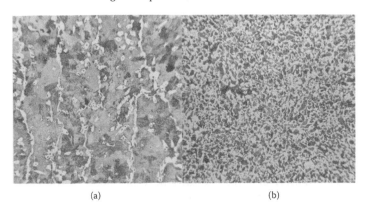

(a)   (b)

FIGURE 3.10
The grain characteristics of weld metal at different locations (a) and (b).

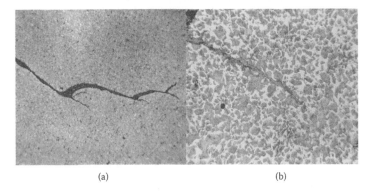

(a)   (b)

FIGURE 3.11
A fatigue crack of approximately 10 mm long and filled with oxides (a) and (b).

3.2.3.3 *Mechanical Testing and Stress Calculation*

In the investigation, the $K_{Ic}(K_{Id})$ test which was needed for stress calculations was not attempted because there was insufficient material. Charpy V-notch tests were conducted instead, firstly to establish whether the parent and weld metal conformed to BS 11 rail steels, and secondly, to obtain the impact values for conversion to K_{Ic} values. The impact values did in fact fall within the range expected of normal BS 11 steel. This finding was further validated by comparing the results from Barsom and Imhof [2] for a series of American rail steels, where the Charpy V-notch impact energies ranged between 3 to 8 ft.lbf (note that 1 ft.lbf equals 1.36 J), and K_{Ic} values ranged from 30 to 45 ksi√in (note that 1 ksi√in = 1.099 MPa√m). Morton, Cannon, Clayton, and Jones have also obtained K_{Ic} values for typical BS 11 rail steels of approximately 30 to 45 ksi√in at 60°C [3]. The results of Charpy testing here suggested that the following values can be reasonably used for calculations:

a. For parent metal: $K_{Ic} \approx 35$ ksi√in
b. For weld metal: $K_{Ic} \approx 40$ ksi√in

Further, for such steels the approximate relationship

$$K_{Ic} = 1.4 K_{Id} \tag{3.1}$$

holds to an accuracy of within 5% [2]. Under dynamic conditions, the dynamic K_{Ic}, designated K_{Id}, is used instead.

For the calculation of stress conditions that cause brittle fracture in a component with a fatigue crack, linear elastic fracture mechanics (LEFM) methods are employed, equating K_I (stress intensity factor) with K_{Ic} (critical value of K_I) and σ (applied stress at infinity) with $σ_{Ic}$ (critical stress or fracture stress); K_I in a component with a flaw under a tensile stress system is defined as

$$K_I = σ\sqrt{(a)}\,\mathrm{Fn}(a, B, W) \tag{3.2}$$

where a is crack size; B and W are specimen dimensions, and Fn(a,B,W) is a function of a, B, W.

The three considerations for stress calculations for the rail are section properties, applied loads and stresses, and critical stresses for vertical wheel loading.

a. *Section properties*: This was calculated by approximating the fracture surface in regions F and G using straight lines to form regular shapes to determine their section areas and moments of inertia (I_{xx}).

b. *Applied loads and stresses*: This was calculated by approximating the stresses that the rail was subjected to under routine operating conditions, that is, thermal, flexural, contact, and residual stresses. Flexural stresses can be further divided into (i) vertical bending moment due to vertical wheel loads; (ii) twisting component due to eccentric vertical wheel loads, and (iii) lateral bending moment due to lateral flange forces. Components (ii) and (iii) were not used in the calculations due to a lack of information; in addition, their actions were probably counteracting. Thermal stresses were estimated using yield strength and the original cross-sectional area of rail and assuming that the force acting upon the rail was truly axial. This also included any force due to rail creep, if present. Resultant stresses would be the difference between flexural stress and the combined thermal and creep stresses, taking into account contractions at night where thermal stress will vanish or reverse direction.

c. *Critical stresses for vertical wheel loading*: This was calculated by estimating the stresses for region F, where stresses would be the highest at the foot of the rail (and progressively reduce towards the rail head). In the absence of a standard solution for K_I, the case is approximated to

$$K_I = Y\sigma\sqrt{(a)} \tag{3.3}$$

From Equation (3.3), the critical stresses at various locations on the fracture surface were estimated by substituting appropriate values of the known parameters. Both static and dynamic conditions were examined.

3.2.4 Discussion

3.2.4.1 *Validity of the Results*

LEFM is an established method and standard procedure for the calculation of critical stresses with an acceptable accrued level of errors. Main errors are expected from uncertainties in obtaining applied forces. However, because of the complex nature of the force system, accurate values could only be obtained from actual measurements. This was impracticable in the present case. Results presented were considered as fair representation of the two extreme limits of the stress conditions suffered by the rail. This was sufficient to eliminate certain probabilities for the fracture at TWJ. It was decided that the critical stresses were significantly lower than the yield strength of the material (462 MPa). So under this condition it can be deduced that the rail will not deform plastically before the failure. This was confirmed by observing the straight length of the rail lying south of the fractured TWJ.

3.2.4.2 Stress Conditions Leading to the Fracture of a Thermit Welded Joint

It was found that the critical stresses at the crack front bound by B at the web (Figure 3.7), were significantly lower than the critical stresses at which a brittle fracture would occur, irrespective of whether dynamic conditions were present or not. Under these conditions, a brittle fracture was not expected to occur. Under static conditions, applied stress at crack front C at the foot (Figure 3.7) approached that of critical stress. However, under dynamic conditions the applied stress is considerably larger than the critical stress, hence a brittle fracture would occur. These conditions were considered to be the two extremes under which the rail would normally operate during the day. The actual operating conditions were assumed to be somewhere between these two extremities, and as such it was sufficient to initiate brittle fracture at the foot at C. Once the foot has broken through, the stresses at crack front B would be increased sufficiently to initiate propagation and to lead to fracture of the web and head. During the night, applied stresses were higher, equivalent to the magnitude of the released thermal compressive stresses. Consequently, conditions at night were considered to be more hazardous than during the day.

3.2.5 Key Conclusions

Results of this investigation revealed that the TWJ originally contained welding defects which had gone undetected over time. Operating stresses caused fatigue cracks to initiate and propagate over time to large sizes, but they remained undetected. When the applied stress levels exceeded the critical stress levels, brittle fractures occurred, first at the foot and subsequently at the web and head, to cause total failure. The calculated applied stresses had exceeded calculated critical stress significantly, so fracture should have occurred a long time ago, yet that had not happened. This scenario suggests that the lateral stresses had been substantive yet subtractive, thus mitigating the high levels of stress suffered by the vertical loads and delaying the onset of fracture.

3.3 Case Study 2: Port Arm Problem in a (Ro-Ro) Ramp

3.3.1 Background

In the late 1990s, a wharf ro-ro ramp in a port in Malaysia broke down during operation. The investigation team observed that the front end of the ramp facing the sea had tilted and the port side had submerged in the water. The port hydraulic piston arm was completely detached from the female nut

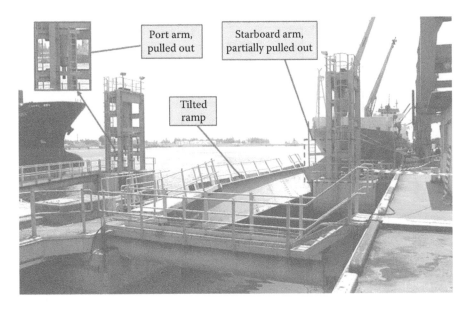

FIGURE 3.12
(See Colour Insert.) Photograph showing the tilted ramp with its port side submerged. The port hydraulic piston arm had pulled out completely from its securing nut; the starboard arm had partially pulled out.

while the starboard arm was partially intact (Figure 3.12). Plainly, failure of the ramp was caused by dismembering of the port arm.

3.3.2 Method of Investigation

Documents on the design of the hydraulic lifting arm were quite scarce. An old manuscript from 1985 (Figure 3.13) was the most useful, claiming that the maximum sustainable load of one cylinder was 260 T.

3.3.2.1 Macro-Examination

3.3.2.1.1 Starboard Arm

For record purposes, when standing on the ramp and looking towards the sea, the right side is called the *starboard* side, and the left side is called the *port* side. Figure 3.14 shows the threaded bottom end of the piston partially removed from the hole of the nut. The nut and its supporting structure had tilted towards the port side. A closer inspection revealed that the threads had stripped off from the female nut and some of them were stuck to the male threads (Figure 3.15). The threads had broken off and there were various crisscross markings on them. The remaining male threads were covered with dark dried grease. Beneath the grease layer, the material was not damaged. These features indicated that damage at this arm were subsequent events.

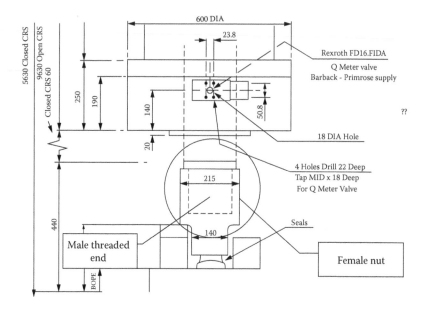

FIGURE 3.13
A side view of the piston arm; the circle encloses the threaded joint that had pulled out at the port side.

3.3.2.1.2 Port Arm

Figure 3.16 shows the port arm with the threaded male end completely removed from the female nut. The female nut had submerged already, so it could not be examined. Figure 3.16c shows the two halves of the male threads, the bottom half (B) with 16 full threads and the top half (T) with 15 full threads. The top half had no female threads and it was covered with some dark, dry grease. The metal at the male end appeared bright and uncorroded, and was probably made of stainless steel or chromium-plated steel. Some female threads were stuck-on at various locations on the bottom half. The stripped-off threads were not continuous and had variable cross sections. At a location about 90° from Figure 3.16, at Figure 3.17, the stripped-off threads covered less than 50% of the total area. In other locations, the male threads indicated no obvious visual signs of corrosion or thinning. Furthermore, 70% of the top and bottom halves were free of female threads on the surface areas.

3.3.2.1.3 Female Nut

The female nut was recovered at a later date. Some of the photographs supplied by the port authority (Figure 3.18) showed three things: the sides of the hole were covered with rust, the male part was completely detached from the hole, and stones and debris had accumulated at the bottom. One part of the thin side of the wall around the hole was removed for examination. Figure 3.19

(a)   (b)

FIGURE 3.14

(a) Starboard arm, viewed from the shore; the structure holding the securing nut had tilted towards the port side. (b) Viewed from the port side, showing the male threads at the bottom end of the arm partially pulled out from the female hole of the nut.

FIGURE 3.15

Close-up view of the male threads; OD of the thread is approximately 160 mm. Threads that had stripped off from the female nut were stuck onto the male threads. The stripped-off threads were not continuous and had variable cross sections. Male threads not containing stuck-on material were darkish in colour due to dried up grease.

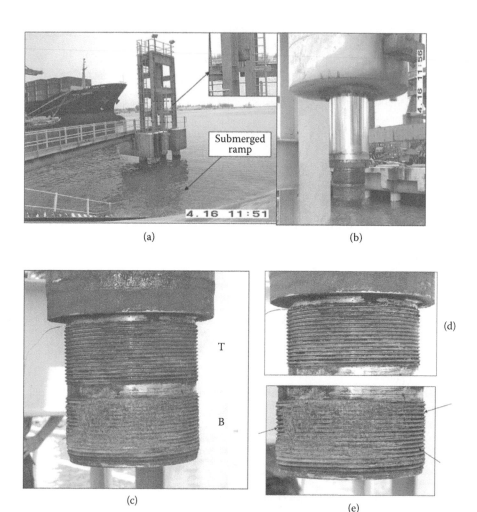

FIGURE 3.16

(a) Port arm; note that the ramp at this side had been fully submerged. (b) Port arm; close-up view of the pulled out male end. (c) Port arm; the male threads are divided into two parts; the bottom half (B) has 16 full threads and the top half (T) has 15 full threads. It was seen that the top half did not have any threads from the female nut stuck on. Port arm male threads; the bottom half has 16 full threads and the top half has 15 full threads. It was seen that the top half did not have any threads from the female nut stuck on. (d) A clearer view of the top half from (c). (e) A clearer view of the bottom half from (c); showing the presence of large amounts of stripped-off threads from the female nut. There were also areas which did not contain stuck-on threads (arrowed). The stripped-off threads were not continuous.

(a)

(b)

FIGURE 3.17
(See Colour Insert.) (a) Port arm; upper part of the male thread, at a location about 90° from Figure 3.16, but with similar characteristics. (b) Port arm; lower part of the male thread, at a location about 90° from Figure 3.16, stuck-on threads covered less than 50% of the area. The cross-sectional areas of the stripped-off threads were variable.

shows a photograph of the surface of the hole covered with rust. The rust on the corroded surface could be easily cleaned off by brushing with a fiber brush with soap. This showed that extra corrosion caused by immersion in the seawater had not been severe. Parts of the thread without mechanical damage were heavily corroded and no longer retained their original profiles, but this corrosion had occurred before the failure. A lengthwise section about 25 mm thick was removed from the sample. From it, the following four specimens were extracted for analysis (Figure 3.20):

a. *Specimen T1*: Located at the top half, with corroded threads having lost their original shape.

b. *Specimen T2*: Located at the top half, containing smeared-over threads.

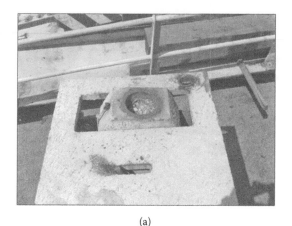

(a)

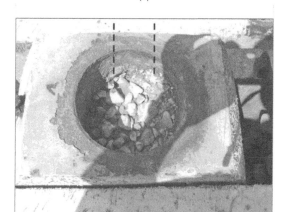

(b)

FIGURE 3.18

(a) Port side; recovered female nut. (b) View of the threaded hole; sides of the hole are covered with rust but it could be seen that the male end had pulled cleanly out; stones and debris had accumulated at the bottom. The thin side at the top of the figure was cut out for examination, along the dotted lines.

c. *Specimen T3*: Located at the top half, with similar characteristics to T1.

d. *Specimen B1*: Located at the bottom half and containing sheared-off threads.

3.3.2.2 Metallographic Examination

3.3.2.2.1 Basic Microstructure and Mechanical Properties

Results of the examination showed that the parent microstructure had 30% of pearlite (darkish constituent) in a matrix of ferrite (light constituent), both equiaxed in nature as seen in Figure 3.21. This indicates that the steel was

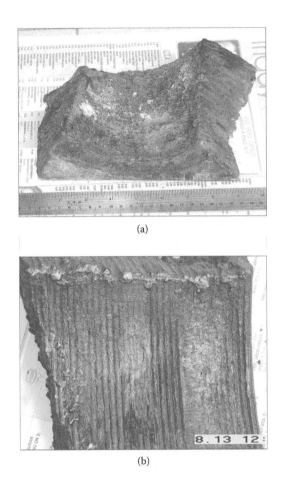

(a)

(b)

FIGURE 3.19
(a) **(See Colour Insert.)** View of a sample cut out from the recovered nut; the surface of the hole is covered with rust. (b) Close-up view of the threaded surface, showing areas with and without mechanical damage on the surface. The threads at locations without mechanical damage had been badly corroded and no longer retained their original profile.

in an annealed or normalized condition. The grain sizes were very coarse. Hardness tests results on the micro-specimens ranged between 148 and 157 HV10, with an average of 150 HV10, corresponding to a Brinell hardness value of 145 HB.

3.3.2.2.2 Specimens T1, T2, T3, and B1

  i. *Specimen T1:* The most complete thread found in this specimen is shown in Figure 3.22. This thread was about 2.1 mm deep but the others were less deep, at about 1.2 mm. All the threads had deep pockets of corrosion on the surface, but had no signs of deformation in the material (Figure 3.23).

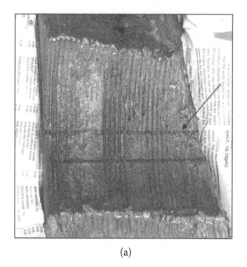

(a)

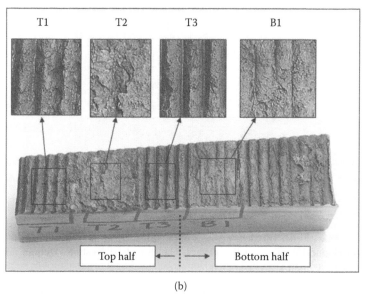

(b)

FIGURE 3.20
(a) The arrows point to a 25-mm thick part sawn out for metallographic examination. This part contains both original and stripped-off threads. (b) View of the sawn out part, showing locations of specimens T1, T2, T3, and B1 that were extracted for metallographic examination. Insets T1 and T3 show heavily corroded threads losing their original shapes. Inset T2 shows threads which had smeared over. Inset B1 shows threads which had been sheared off.

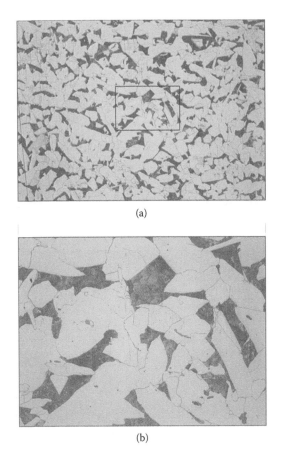

(a)

(b)

FIGURE 3.21
(a) At 50× magnification, the parent microstructure, consisting of about 30% of pearlite (dark-ish constituent) in a matrix of ferrite (light constituent), both equiaxed. The grain size is very coarse. (b) 200× magnified view of the central rectangle from (a).

   ii. *Specimen T2*: Figure 3.24 shows a thread with a partially truncated crown, a typical microscopic appearance of the threads found at the smudged sites. The truncated surface had no local deformation but some fairly advanced corrosion.

   iii. *Specimen T3*: Figure 3.25 shows a thread with the crown partially truncated. No obvious signs of local deformation were found but the whole thread was bending slightly towards one side and the surface area was badly corroded.

     In short, the remaining threads at the top half of the nut were either badly corroded or truncated. Fairly advanced corrosion in the truncated surfaces suggests that they were not caused recently. The

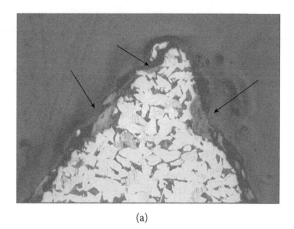

(a)

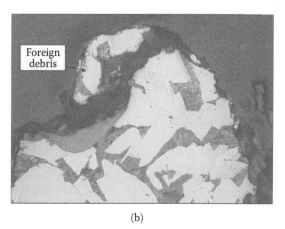

Foreign
debris

(b)

FIGURE 3.22
(a) **(See Colour Insert.)** At 50× magnification, specimen T1, tooth No. 2 was one of the most complete threads at T1, measured to be 2.2 mm deep from the crown to the root. The arrows point to deep pockets of corrosion on the surface. No signs of deformation were present in the material. (b) 200× magnified view of the tip of the crown, showing a lack of deformation in the material. The loose piece is a bit of foreign debris.

truncated surfaces had no local deformation so truncation must have been caused by low-stress or low-energy processes. This suggests that the threads had been largely ineffective at the time of failure.

iv. *Specimen B1*: Figure 3.26 is an illustration of a typical sheared-off thread with little corrosion on the sheared surface, unlike the top parts T1, T2, and T3. This indicates a freshly formed surface, that most likely resulted from the failure incident. These threads had developed their full capacity before failing in a ductile manner and were very likely the ones that were supporting the piston at the time of failure.

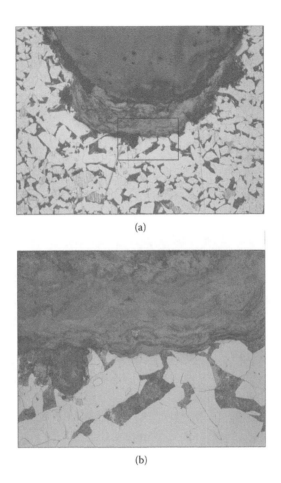

(a)

(b)

FIGURE 3.23
(a) At 50× magnification, specimen T1, the thread root between the second and third threads; the depth of corrosion was about 0.6 mm. (b) 200× magnified view of the central rectangle from (a).

3.3.3 Discussion

3.3.3.1 Nature and Direct Cause of Failure

The investigation found that the front port side of the ramp had dropped into the sea and submerged. The ramp failed because the male end of the hydraulic arm piston at the port side had dislodged from the female nut. Subsequently, the hydraulic arm at the starboard end also pulled out.

3.3.3.2 Materials of Construction

The male end of the port and starboard pistons were either made of stainless steel or chromium-plated steel since they did not show any visual signs of

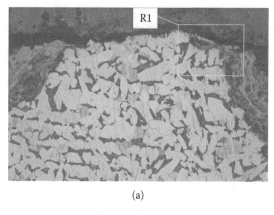

(a)

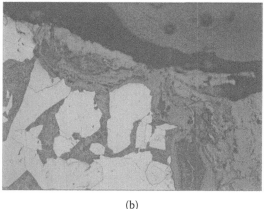

(b)

FIGURE 3.24

(a) At 50× magnification, specimen T2, the thread with a partially truncated crown; the truncated surface does not show any obvious signs of local deformation. (b) 200× magnified view of rectangle R1 from (a), showing a lack of observable deformation at the truncated surface; the surface also contains fairly advanced corrosion products.

corrosion. The male threads were made of stronger materials than the plain carbon steel in the female nut. Plain carbon steel conforms to the mechanical properties of BS 4360 Grade 43 and judging from the thickness and coarse grain size, its yield strength was estimated to be 225 MPa.

3.3.3.3 Root Cause of Failure

The port male end threads could be easily dislodged because the female nut did not have sufficient sturdy threads to secure it fully. Probably only 30% of the 31 threads were effective. At the starboard side, apparently there were more good threads but no count was made since failure at that end was a

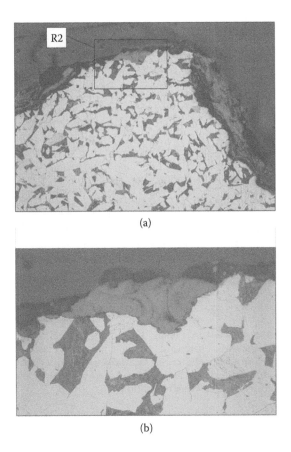

(a)

(b)

FIGURE 3.25
(a) At 50× magnification, specimen T3, the thread with a partially truncated crown; the truncated surface does not show any obvious signs of local deformation but the whole thread appeared to have bent slightly towards the right of the figure. (b) 200× magnified view of rectangle R2 from (a), showing a lack of local deformation at the truncated surface; the surface also contains fairly advanced corrosion products.

subsequent event. It was thus concluded that more threads at the port side had become ineffective because of the following reasons:

a. Corrosion wastage at the threads, especially at the top half which was more exposed to the elements.

b. Previous loss of material (truncation) at the crown of the threads due to unknown events.

Degradation of the female threads was caused by electrochemical corrosion from moisture ingress in the spaces between the male and female threads.

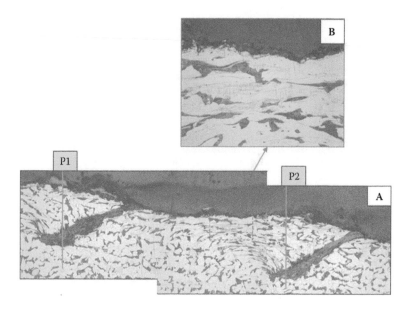

FIGURE 3.26
(See Colour Insert.) Specimen B1: (A) Planes P1 and P2 show the locations of adjacent roots of a thread; the thread has sheared off, towards the right of the figure and caused the underlying material to become deformed. The sheared surface did not contain deep pockets of corrosion at 25× magnification. (B) Magnified view of the sheared surface, showing deformed ferrite (light) and pearlite (darkish) grains at 100× magnification.

The use of two different steels for making the nuts in the male end, namely, a weaker carbon steel and a stronger stainless/chromium-plated steel was a mistake. The galvanic action between the two materials would have accelerated corrosion at the less noble nut material.

3.3.3.4 Remaining Nut Capacity

The calculations displayed in Appendix 3.1 show that the nut, if functioning as designed, would have a capacity of 6898 kN and a safety factor of 2.70 over the maximum sustained load of 2551 kN. However, if only 30% of the threads were effective as estimated, the nut capacity would have only been 2069 kN or 81% of the maximum design load of 2551 kN. In engineering parlance, this is a very unsafe level indeed.

3.3.4 Conclusion

To conclude, the investigation revealed that the failure of the wharf ro-ro ramp happened when the threaded end of the port side piston pulled out from the mating female nut. This occurred because barely one-third of the female nut threads were effective. Extensive corrosion of the threads in the female nut, especially at its top end, had reduced its capacity to only 81% of its maximum design load of 2551 kN.

3.4 Case Study 3: Collapse of a Girder Launcher

3.4.1 Background

A girder launcher was used to erect a 40-meter span for Kuala Lumpur's light rail transit system in the late 1990s. The launcher comprised two parallel girders, called the left and right girders when looking in the direction of travel. After the launcher had erected some of the planned segments it suddenly collapsed. Some workers on top of the launcher fell but luckily they were not badly hurt. Fortunately, the construction was in a remote area so the launcher accident did not cause any property damage or traffic disruption. The girders were fabricated overseas but underwent some repairs locally before being put into operation.

3.4.2 Method of Investigation

3.4.2.1 Site Examination

An examination was initially carried out at the site of the collapse. Later, the hinge sections were extracted and cleaned to see if there were problems with the welding. The examination was generally confined to the parts directly involved in the failure. There were many other signs of unsatisfactory welding but since they were not directly related to the failure mechanism, they will not be given prominence here.

3.4.2.1.1 General Views of the Collapsed Launcher

Some general views of the girder launcher are shown in Figures 3.27 and 3.28. The photographs indicate that the hinges of both the left and right girders had become separated at the joints between the male hinge plates and the stiffener boxes, making the central portions fall to the ground. The whipping force from the sudden impact bent the two nose elements at each extreme of the girder, downwards, at their hinged joints. In their final positions, the broken central part of the left girder rested on an embankment adjacent to the forest whilst the right girder rested on the ground. Both girders were rotated inwards facing each other, from the weight of the concrete segments placed in between. The portal crane that was depositing one of the segments at the front end had rolled backwards and downwards into the central portion during the collapse.

FIGURE 3.27
(See Colour Insert.) The right girder, showing the broken, central, hinged portion resting on the ground, and the ends still supported on the pier brackets.

FIGURE 3.28
The left girder, after lifting and placing it on the embankment. The launching direction was towards the right of the figure. Note the two torn-off male hinge plates attached to the female plates.

FIGURE 3.29
Left girder, resting on an embankment, as seen from the outward side next to the jungle.

FIGURE 3.30
The right girder, stiffener box connection of the upper male hinge plate; all fractures had occurred at the weld metal and had a bright appearance.

3.4.2.1.2 Visual Examination

Figure 3.29 shows the broken ends of the left girder that had become partially buried in the embankment, restricting access for examination. However, Figure 3.30 of the right girder, shows the male, stiffened connection box with the male hinge plate torn off from the weld joints; all fractures had a bright appearance. Figure 3.28 also shows the left girder after it was removed and placed on the embankment; this part shows the female hinges with the

FIGURE 3.31
(See Colour Insert.) The left girder, stiffened connection box of the male hinge plate; 'BF' stands for bottom flange.

torn-off male hinge plates still pinned to them. The lower connection of the male hinge plate (Figure 3.31) was removed for inspection. The following faults were found:

1. The primary failures had all occurred because of tension and shear forces at the weld metal connecting the male hinge plate to the stiffener box. The weld thicknesses/throats at the fracture planes were much thinner than the specified values, ranging between 0 and 12 mm, averaging less than 5 mm. The actual specifications called for full penetration welds. The weld areas of all the joints were estimated and compared with the specifications listed in Table 3.1, and found to be grossly deficient.

2. There were no cleavage characteristics on all of the fracture surfaces, showing that none of the fractures were brittle.

3. All fractures were ductile in nature with predominantly slanted directions typical of shearing mechanisms.

4. Plate G244 was supposed to be one continuous piece of metal, but cuts were found at the positions shown in Figure 3.33, but apparently, this condition had not been subject to a weld repair. G247 should have been fused to G244 and G245 but this had not been done.

5. Weld preparation was inadequate or absent in all the broken joints (see the 'P' arrows in Figure 3.33).

6. Foreign material was occasionally used to fill up the gaps between the plates, as shown in Figures 3.33 and 3.34.

TABLE 3.1

Designed Values and Actual Values

Weld Joint	Designed Area (mm²)	Actual Area (mm²)
G245/G240	25,360 (s)	2970 (s)
G245/G242	25,360 (s)	4290 (s)
G245/G246	13,440 (t)	4013 (t)
G245/G247 and G82	5600 (t)	2064 (t)
G245/G243	N.A[1]	1530 (t)
Total equivalent tensile area[2]	48,323	11,798[3]

Note: (s) = Area under shear stress; (t) = Area under tensile stress.
[1] Not applicable: welds made according to specifications will not fail along this path.
[2] Calculated by dividing the shear area by 3.5 and adding to the tensile area.
[3] Actual stress area was only 24.4% of the designed area.

FIGURE 3.32
The left girder, lower connection hinge with the male hinge plate and welded-on parts, still pinned to the female plates.

7. The plates marked 'X' in Figure 3.35 had been previously cut out and subsequently replaced but the plate preparation and weld penetration still fell short of the specified requirements.

8. The fracture at the joint (arrow 3 in Figure 3.36) between hinge plate G245 and vertical plate G244 had occurred on parent metal G244. The fracture had a step appearance similar to one caused by lamellar tearing. However, from the manner of the deformation in G244, it was

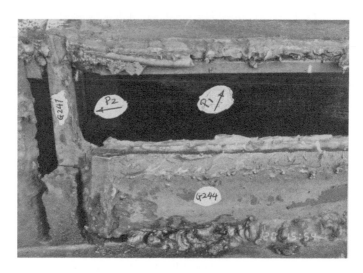

FIGURE 3.33
The left girder, stiffener connection box to which the male plate of Figure 3.32 is attached. Note the lack of penetration at the top weld, and the presence of the piece of included foreign metal (probably a welding electrode). Note also the generally poor welding quality (bottom weld).

FIGURE 3.34
The left girder, male hinge plate of the lower connection, as seen from the outward side.

FIGURE 3.35
The left girder, stiffener box of the lower connection, with G240 and the bottom flange removed. 'X' indicates the areas previously cut out and replaced. The arrow points to an abnormal looking weld between G244 and the bottom flange. The white dashed lines indicate the boundaries of the sample to be cut out for examination.

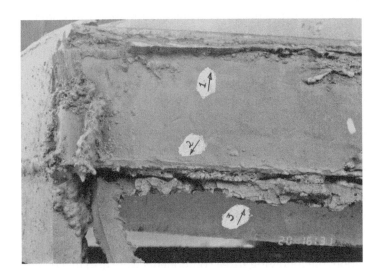

FIGURE 3.36
The left girder, male hinge plate of the lower connection, as seen from the front, and showing a view of the inward edge. Arrow '1' points to the fractured weld metal between the 80-mm thick G245 and the 20-mm thick G246; this was an original weld. Arrow '2' points to the fractured weld metal between G245 and G244. Arrow '3' points to metal which had torn off from G244; the fracture surface at this part had a stepped nature. It can be seen that the weld thicknesses at the fractures were well below the specified values of 20 mm.

deduced that this is a secondary fracture which had happened after G245 and the stiffener box became detached.

9. The weld between plate G244 and the bottom flange had a shockingly bad profile (see Figure 3.33 and Figure 3.35). It appeared as if they had applied the weld after hinge plate G245 was fitted so penetration or bonding was practically impossible.

10. Plate thicknesses were all found to conform to the specified design.

In addition to the observations outlined in items 1–10 above, the following also occurred in the various components:

- *Upper connection, male hinge.* Failure at the weld metal connecting the male hinge plate to the stiffener box due to inadequate weld preparation and lack of weld metal, similar to the situation found in the lower connection. There were no signs of failure due to compression forces, which would have caused buckling or crushing. Since it did not collapse because of compression, the male hinge in the upper connection could not have been the first to fail. So the matter was disregarded.

- *Female hinges.* Both the female hinges were still intact so it could not be established visually whether similar inadequacies existed. Since no failure occurred here, no further examination was undertaken.

- *Right girder, main hinge.* This girder failed subsequent to the left girder, as reported by an eyewitness. The mode of failure in this hinge and the welding inadequacies were practically identical to those in the left girder. Only the failure path of the male lower connection was slightly different (compare the differences shown in Figures 3.37 and 3.38). However, the difference is not significant. The female, lower connection was forced upwards when the girder hit the ground (Figure 3.38). Some of the welds connecting the hinge plates and stiffener box to the bottom flange had broken. The quality of the welds was as bad as that used elsewhere. When the bottom flange beneath the binge plate was extracted and examined, the specified weld between the 20-mm thick G368, the bottom flange G46, and the female hinge plate, were missing (Figures 3.39 and 3.40).

In addition, the following three components were also inspected:

- *Sliding bearing supports.* Inspection of the bearing supports on site and after removal showed that they had been adequately secured. All linkages failed in a ductile manner as a result of extreme stress.

- *Clamping plates and bolts.* Clamping plates were placed on the outer bearing to prevent the girder from toppling inwards when the

FIGURE 3.37
The left girder, male hinge plate of the lower connection, as seen from the inward side. Very little weld metal existed between hinge plate G245 and plate G240 of the stiffener box (not shown), and between vertical plate G242 and the bottom flange.

FIGURE 3.38
(See Colour Insert.) The right girder, lower connection, with the torn-off male hinge plate attached to the female parts. The female members had been forced up on hitting the ground.

segments were installed on the element support beams. In this case, some plates had been torn off when their clamping bolts broke or were torn off. All such bolts had either failed by ductile mechanisms due to gross overloading, or their threads had stripped off a substantial length. The bolts were all M24 Grade 8.8. Laboratory hardness

FIGURE 3.39
The right girder, bottom flange (G46) of the female hinge, after cutting out; the arrow points to G368, which was supposed to connect the bottom plate to female hinge plate G215. There were no signs that there had ever been a weld between G368 and G215.

FIGURE 3.40
Magnified view of G368, showing the lack of any weld between it and G245.

tests on four randomly chosen samples gave a hardness value of between 32.4 and 43.6 HRC on the bolt-heads, which exceeded the Grade 8.8 bolt specification.

- *Curve regulation cylinders.* These components are shown in Figure 3.41. They were reported to be in good working order as shown in

FIGURE 3.41
The right girder, magnified view of the separated hinge; the torn-off male hinge plate is on the left part. Note also the torn-out curve regulation cylinders.

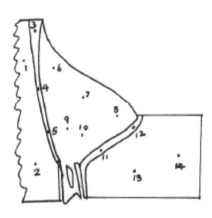

	Specimen No. 2
Point	HV10
1	185
2	175
3	213
4	203
5	222
6	209
7	202
8	197
9	171
10	178
11	207
12	198
13	187
14	191

FIGURE 3.42
Hardness traverse on micro sections.

Figure 3.42. Accordingly, examination by the investigator/writer showed that all the locking nuts were approximately in the same position. No further examination of these three components was considered to be necessary as they were minor factors, and these faults would not have contributed to the ultimate collapse of the launcher.

3.4.2.2 Laboratory Examination

Laboratory examinations were performed mainly to evaluate quality of the materials used, quality of the welding, and fracture mechanisms. Only optical microscopy was used because information obtained therein was considered to be fully sufficient for the purpose at hand.

3.4.2.2.1 Sample Extraction

The part of the box section enclosed by the dotted lines (Figure 3.35) was the piece cut out and transported for examination in a metallurgy laboratory. At the laboratory, the cut-out box was sectioned into two parts followed later by more dissections later.

- *Visual examination and fractography using a stereo-microscope*: All the failures proved to be ductile in nature, and they all appeared slanted (except for those in G245 and G246). There were no markings on the fracture surfaces to indicate which particular weld had failed first, much less the fracture origin at any fractured joint. These were not expected in failures of this nature.

- *Macro-examination (metallographic)*: Some features are shown in Figure 3.41 onward.

  a. *Figure 3.43, Specimen 1*: The original weld did not have the proper preparation, resulting in a gross lack of penetration. A second weld was laid on top, apparently in an effort to strengthen the first weld.

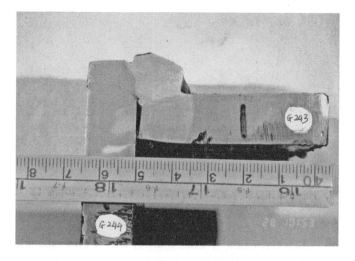

FIGURE 3.43
Specimen 1, etched macro-view; showing that the original weld had less than 5 mm of penetration. The weld had obviously not been beveled for full penetration. The excessively convex profile had probably been due to building up by a second contractor.

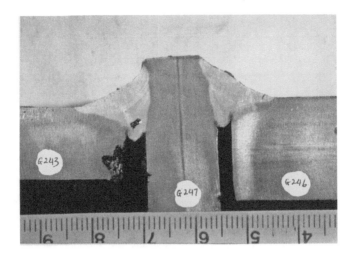

FIGURE 3.44
Specimen 2, etched macro-view; showing the original welds with highly inadequate penetration. The welds had also obviously not been beveled for full penetration. The central dark zone G247 is due to heavy segregation in the steel.

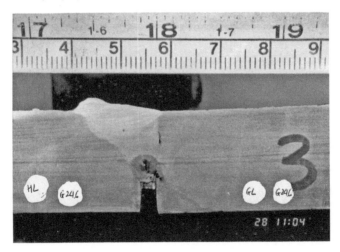

FIGURE 3.45
Specimen 3, etched macro-view of weld in G246; inadequate preparation and penetration, a defective root run and a lack of fusion.

    b. *Figure 3.44, Specimen 2*: The original weld with poor preparation and a gross lack of penetration.

    c. *Figure 3.45, Specimen 3*: A weld with inadequate preparation and penetration, a defective root run and a lack of fusion.

    d. *Figure 3.46, Specimen 3*: A broken joint between G245 and G246— Same welding condition as in item c (above).

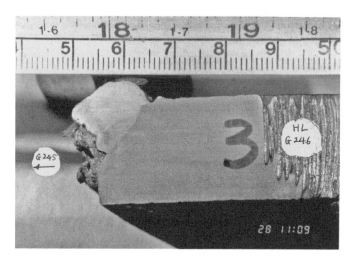

FIGURE 3.46
Specimen 3, etched macro-view of fractured weld between G246 and G245. Penetration was about 50% of the plate thickness. Note the slant nature of the fracture.

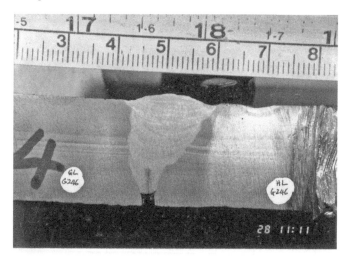

FIGURE 3.47
Specimen 4, etched macro-view of weld in G246; almost full penetration had been achieved, but hot cracking was present at the root run.

    e. *Figure 3.47, Specimen 4*: The welding had been made by a second contractor. Almost full penetration was achieved but a hot crack was present at the root run.

    f. *Figure 3.48, Specimen 4*: A broken joint between G245 and G246— The small piece 'Z' was part of the original plate that a second contractor had not removed when they opened up G246. A small

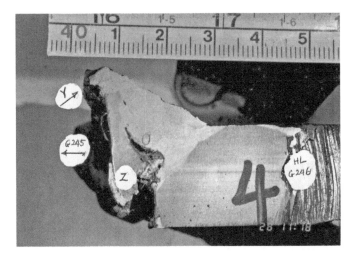

FIGURE 3.48
Specimen 4, etched macro-view of the fractured weld between G246 and G245; part 'Z' is a small part of the original G246 not removed. Arrow 'Y' points to the highly deficient weld; the adjacent weld to this was more substantial, though still inadequate. Note the slant nature of the fracture.

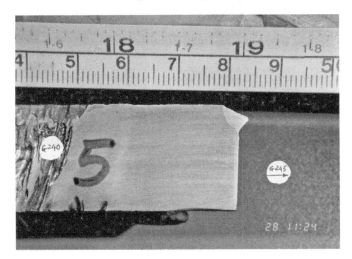

FIGURE 3.49
Specimen 5, etched macro-view of fractured weld between G240 and G245; the fractured weld throat was only about 3 mm thick. Fracture had been by shear.

part of the weld 'Y' was the original weld joining G245 to G246. The original weld was not beveled but the second weld was, to 10 mm deep. Contrary to the specification both welds did not fully penetrate the joints.

g. *Figure 3.49, Specimen 5*: A highly deficient original weld.

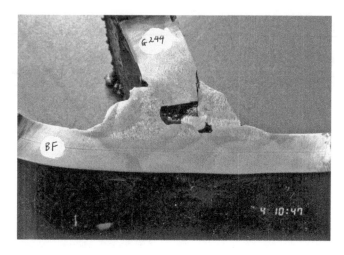

FIGURE 3.50
Specimen 6, etched macro-view of the distressed joint between G244 and the bottom flange; welds on both sides had been built up.

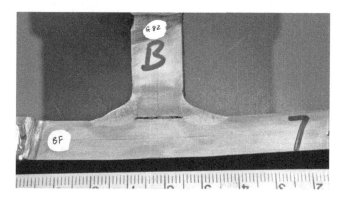

FIGURE 3.51
Specimen 7, deficient fillet welds between G82 and the bottom flange.

   h. *Figure 3.50, Specimen 6*: A distressed joint between G244 and the bottom flange; welds had been built up by a second contractor.

   i. *Figure 3.51, Specimen 7*: Deficient fillet welds between G82 and the bottom flange.

- Metallographic examination and hardness test: For this purpose, sections were extracted from Specimens 2, 3, 4, 5, 6, and 7. Care was exercised to ensure that they were taken at least 20 mm away from the flame-cut edges to avoid the heat-degraded microstructures. Hardness traverse values are presented in Figure 3.42. Results showed that both the parent and weld metals conformed to the material specifications. Tensile tests also showed that the material conformed to specifications.

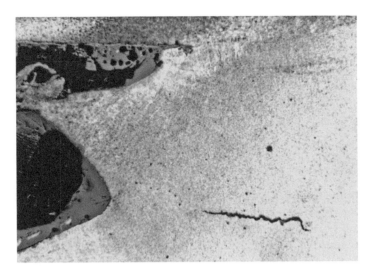

FIGURE 3.52
Specimen 2, etched microstructure of a weld between G247 and G246; showing a small hot crack at the root run.

- Results of the metallographic examination:

    1. All the plates were banded to varying degrees. Some contained excessively large amounts of elongated manganese sulphide (MnS) stringers and other forms of non-metallic inclusions.
    2. In G244, the step appearance of the torn-off metal (see arrow 3 in Figure 3.36) was caused by tearing along MnS stringers, a process similar to lamellar tearing.
    3. Quality of weld metal was generally normal. However, hot cracks were present in both the original welds and the welds made by another contractor (Figure 3.52).
    4. Fracture surfaces from Specimens 3, 4, and 5 had distinct shear characteristics.

- Stress at failure: After the collapse of the launcher, the actual stresses borne by the hinge connection were calculated by another consultant, to give:

Maximum moment at the hinge:	15.119.5 kN.m
Distance between male hinge plates:	1914 mm
Maximum tensile force at each connection:	7899.4 kN
Specified equivalent tensile area:	48,323 mm²
Expected working stress at the lower male hinge joint:	163 MPa

However, as a result of the deficient weld area, the actual average stress experienced by the weld joint had increased to approximately 670 MPa. This stress number is obtained by dividing the maximum tensile force given above by the measured total equivalent tensile area in Table 3.1. Note that the maximum moment at the hinge as shown in the above case was only 72.5% of the worst-case loading design.

3.4.3 Discussion

3.4.3.1 Nature and Sequence of the Failure

Failure of the launcher was precipitated by the separation of both the lower and upper connections of the left and right girders. When these occurred, the central portions became free and collapsed. Separation of the four connections was all caused by the male hinge plates tearing off from the stiffener boxes; the two components should have been welded together firmly. Examination of the collapsed girders could not show which connection failed first; however, eyewitness accounts identified the left girder to have collapsed before the right one. Since the girder was subjected to sagging moments and since the upper connection did not fail because of compression forces, it was concluded that it was the lower connection of the left girder which first failed. This was followed by failure of the left, top connection, and then the right girder.

3.4.3.2 Primary Cause of the Failure

In all four connections, fracture had occurred at the weld metal of the joints by ductile failure. In all the failed joints, the thickness of the weld metal was grossly deficient, and this was identified as the primary cause of failure. It was impossible to determine which joint had failed first, or if at all, a few had failed simultaneously simply because there were no telltale marks made on the fractured surfaces, which would be present in a brittle or fatigue failure. However, this did not detract from the usefulness of the report to the client and to the Department of Safety and Health (DOSH).

3.4.3.3 Design of the Girder and Hinge

The general design of the launcher was certified as safe after the incident by third-party structural engineers, though certain localized design weaknesses were detected, which had to be rectified. The failure itself, if taken as a full-load test, showed that only 24.4% of the weld metal was able to carry out some launching operations before failure occurred. But of course, such structures, by regulation, must be designed to be absolutely safe under the worst loading conditions possible.

3.4.3.4 Materials Used

The metallurgical and mechanical quality of the plate and weld materials had not been lacking and had not contributed to the failure.

3.4.4 Conclusion

The launcher failed due to grossly deficient weld metal at the welded joints of both girders. The reason for this deficiency was not investigated as it was outside the purview of the Failure Investigator, but it is clear that the overseas fabricator had not done an honest job.

Problems and Answers

Problem 3.1

What are the key features of macro- and metallographic examination employed commonly in failure analysis?

Answer 3.1

1. Macro-examination is commonly conducted to inspect:

   (a) The physical attributes of the damage suffered
   (b) Condition (including defects) of the specimen/parts
   (c) The appearance of the fracture surfaces, the crack path(s), and possible fracture mechanisms

   Metallographic examination is commonly conducted to establish the microstructures of the materials involved for evidence of abnormalities and for details of failure modes.

Problem 3.2

What are the major assumptions made when employing linear elastic fracture mechanics?

Answer 3.2

In LEFM, it is assumed that:

- Crack has been initiated
- Material is linearly elastic

- Material is isotropic
- Crack has started to propagate
- Plastic zone near crack is small
- Points of analysis are near the crack tip

Problem 3.3

In the case of the rail track failure in Case Study 1, what was the purpose of the Charpy V-notch tests?

Answer 3.3

(a) To establish whether the parent and weld metal at the weld joint exhibited fracture behavior similar to that of the normal BS 11 rail steel.

(b) To obtain impact values for conversion to K_{IC} values, in order that LEFM calculations may be made.

Problem 3.4

What is the effect of corrosion upon the threads on the hydraulic arm piston in the case of the ro-ro ramp in Case Study 2?

Answer 3.4

The effects of corrosion are:

(a) Wastage at the threads
(b) Truncation at the crown of the threads
(c) Factors (a) and (b) above which had caused total loss of function in the threads

Problem 3.5

How can you ensure that the weld in a structural component is of adequate quality?

Answer 3.5

By following the provisions of an approved welding code, such as the American Welding Society (AWS) D01.1. Such welding requires proper supervision, qualified welding procedures, qualified welders, possible preheat and postheat operations, and various levels of non-destructive testing.

Appendix 3.1: Calculations on the Capacity of the Nut

1.0 Basic Specifications

(a)	Maximum designed sustained force:	260 T (2551 kN).
(b)	Outer diameter of male thread (measured by KPC):	160 mm
	Pitch diameter (calculated):	157.4 mm
(c)	Thread pitch (measured by writer):	4 mm
	Thread thickness at pitch diameter:	2 mm
(d)	Thread form (deduced from drawings):	metric
(e)	No. of threads at bottom half:	16
	No. of threads at top half:	15
(f)	Hardness of steel:	145 HB
	Estimated tensile strength:	450 MPa
	Estimated yield strength for the nut material:	225 MPa

2.0 Shear Stress Area of the Threads at the Pitch Diameter

Stress area per thread:	989.0 mm^2

3.0 Designed Load Capacity of the Threads

(a)	If all 31 threads were functioning as designed, the total stress area would be:	30,658 mm^2
(b)	Then, total capacity of threads against yielding would be:	6898 kN
(c)	Factor of safety against yielding would be:	2.70

4.0 Estimated Actual Capacity of the Threads

(a) Assuming that only 30% of the threads were functioning, actual capacity of the threads would be: 2069 kN

(b) Factor of safety would be 0.81, which is less than 1

References

1. Frederick, C.O. and E.G. Jones, Review of Rail Research on British Rail, *Transportation Research Record 744*, Washington, DC: National Academy of Sciences, 1980.

2. Barsom, J.M. and E.J. Imhof, Jr., Fatigue and Fracture Behaviour of Carbon-Steel Rails, *Rail Steels—Development, Processing and Use, ASTM STP 644*, American Society of Testing and Materials, 1978.

3. Morton, K., D.F. Cannon, P. Clayton, and E.G. Jones, The Assessment of Rail Steels, *Rail Steels—Developments, Processing and Use, ASTM STP 644*, American Society for Testing and Materials, 1978.

4. Besuner, P.M., Fracture Mechanics Analysis of Rails with Shell-Initiated Transverse Cracks, *Rail Steels—Developments, Processing and Use, ASTM STP 644*, American Society for Testing and Materials, 1978.

4

Mining and Production System

4.1 Introduction

Mining has been carried out since ancient times to extract valuable minerals and other geological materials from the earth. In the pre-1990s, tin mining was a major industry in Malaysia, which was then the largest producer in the world. However, the combination of dwindling deposits, escalating costs, and low prices caused the industry to collapse. In 2007, China and Indonesia became the largest producers, each contributing approximately 43% to the global market. About one-third of all the tin mined in the world today comes from the Indonesian islands of Bangka and Belitung. Tin dredges are best used where there are large expanses of alluvial soils without excessive rock formations and also offshore. Tin dredges are large machines and can weigh more than 4500 tonnes, with a floating support—a pontoon in excess of 75 meters in length and 35 meters in width. The largest dredges can be equipped with up to 150 digging buckets, each weighing more than 3.5 tonnes and digging up to 45 meters in depth. Tin dredging has evolved over a few decades and in its present form, unchanged since the late 1980s, was one of the most efficient methods of mining. Malaysia at that time together with international consultants such as W.F. Payne & Sons, was at the forefront of dredge design and manufacture. Most of the other items were locally cast or fabricated except for a few specialized items such as large motors and gearboxes. Similar to the rest of the world in those days, calculations were made with slide rules, computers had not been invented yet, and design standards were still quite rudimentary. Many components were large in size and were often over-designed and made to less than exact specifications. When field problems or failures were encountered, engineers had to solve them fast to reduce loss of production. They were required to have hands-on experience and the ability to plan and execute repairs quickly and efficiently, and often to local sources that made spares of adequate quality. Recurring problems such as those described in Case Studies 2 and 3 were outsourced to the appropriate specialists for investigation.

Three case studies of failures in mining are discussed in this chapter. The first case is quite current and Case Studies 2 and 3, though of some age, are still relevant where tin dredging is still carried out. The cases are:

Case Study 1: CO_2 Attack on Oil Well Tubing
Case Study 2: Tin Dredge Pinion Wheel Failure
Case Study 3: Excessive Wear of Bucket Pins in a Tin Dredge

Generally, failure analysis of tin mining components follow the procedures described in Chapters 1 and 2.

4.2 Case Study 1: CO_2 Attack on Oil Well Tubing

4.2.1 Background Information

This reports details the failure investigation into the CO_2 attack of short string (SS) and long string (LS) production tubes from an oil well. The SS tubes were said to be severely affected from about 2200 ft MD BDF to about 3550 ft MD BDF. The LS tubing was said to have parted at about 3000 ft MD and to have many holes in the tubing. Product analysis was not known. The tube material was given as API 5CT, Grade L-80 [1].

The FI made an inspection visit to the storage yard in May 2009. Several tube samples were marked out for laboratory examination and these were delivered in due course.

4.2.2 Site Inspection

Both the SS and LS tubes were stacked together, without any method to differentiate them except by eye. Many tubes on the top layer contained severe corrosion damage, which in many cases had breached the tube wall. Significantly, the damage was only found on one side of the tube and was variable in intensity.

A close-up view of damaged tubes (Figure 4.1a) shows highly irregular profiles due to internal pitting attacks. The corroded edges had a darkish-coloured scale (inset), different from the reddish-brown colour of the scale on the unaffected surfaces.

Certain tubes, said to be LS, appeared to have external pits (Figure 4.1b); a close-up view of the pits showed that they contained a darkish scale within the pits (Figure 4.1c), similar to the internal pits (Figure 4.1a).

Many tubes had collars with corrosion attacks on their external surfaces; the attacked areas were invariably dark in colour. The tube sections adjacent to many of the affected collars were also badly corroded, with their walls breached.

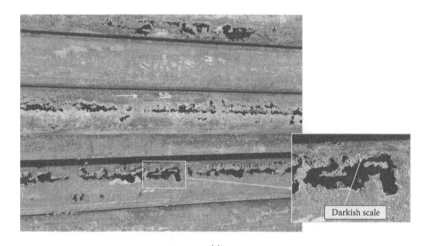

(a)

(b)

FIGURE 4.1
(See Colour Insert.) (a) Close-up view of the damaged tubes, showing highly irregular profiles due to internal pitting attacks. The corroded edges had a darkish-coloured scale (inset), different from the reddish-brown colour of the scale on the unaffected surfaces. (b) Certain tubes appear to have external pits (said to be LS tubes). (c) Close-up view of the external pits from (b): The arrow points to a darkish scale within the pits, but the scale on the unaffected surfaces was reddish brown in colour.

Figure 4.2a,b shows collars with extensive corrosion on their external surface; the attacked areas were sharply demarked from the non-attacked areas, and also had a darkish colouration.

Figure 4.3a through Figure 4.7 show samples marked out for extraction and laboratory examination. The samples were:

- S11, S21, and S31 from the SS tubes.
- L11 and L21 from the LS tubes; L21 contained a corroded collar.

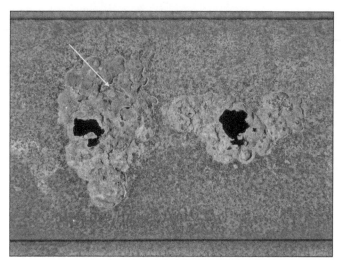

(c)

FIGURE 4.1 (*Continued*)
(See Colour Insert.)

A loose sample created through an internal pitting attack was also examined (Figure. 4.8); the surfaces formed by the corrosion attack were highly irregular and contained darkish deposits.

4.2.3 Summary of the Findings from the Site Inspection

From the above site inspection, the following preliminary views were obtained:

- The corrosion attack had occurred at the SS and LS tubes and at their collars.
- In all cases, the attack was due to pitting, and the pitted areas were sharply separated from the unaffected regions. In the SS tubes, the attack had originated at the internal surface, but for the LS tubes and collar, the definite origin of the pits could not be established.
- The surfaces of all the pits had a darkish scale, quite different from the normal reddish-brown corrosion scale.
- The characteristics of the corrosion damage suggested that it was some form of CO_2 attack.

4.2.4 Visual Examination

The received samples are shown in Figure 4.9; examination of the internal surfaces of samples L11 and L21 showed that severe pitting had also occurred at the internal surface (Figure 4.10a,b and Figure 4.11a,b). The pitted areas also contained a darkish scale.

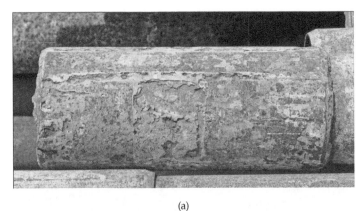

(a)

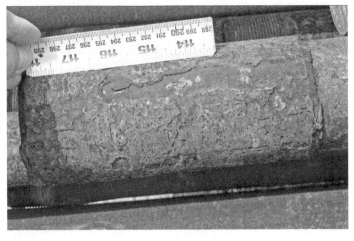

(b)

FIGURE 4.2
(See Colour Insert.) (a) The collar with extensive corrosion on its external surface: Attacked areas are sharply demarked from non-attacked areas, and it also has a darkish colouration. (b) Another collar with extensive corrosion on its external surface: Attacked areas are also sharply demarked from non-attacked areas, and it also has a darkish colouration.

A laboratory testing program was drawn up, as detailed in Table 4.1; Figure 4.9 shows the locations and identity of the cut-out specimens. The purpose of the tests was to establish whether the materials used conformed to specifications and whether the darkish scale was $FeCO_3$, which would indicate a CO_2 attack.

4.2.5 Gas Chromatography

If $FeCO_3$ were present on the specimen, a reaction with HCl would produce CO_2, which would be detected by the chromatograph. Normal oxides of steel would not produce this gas.

(a)

(b)

FIGURE 4.3
(a) Sample S11, from a badly affected SS tube. (b) Close-up view of S11.

FIGURE 4.4
Sample S21 from another badly affected SS tube.

FIGURE 4.5
Sample S31 from a lightly damaged SS tube.

The results presented in Table 4.2 show large amounts of CO_2 [2]; this result is a strong indication for the presence of $FeCO_3$.

4.2.6 Chemical Analysis

The specimens were sent to an accredited laboratory and analyzed using an optical emission spectrometer.

The results presented in Table 4.3 show that the chemical composition of all the specimens tested conformed to API 5CT.

FIGURE 4.6
Sample L11 from a lightly affected LS tube.

FIGURE 4.7
Collar sample L21 from an LS tube, with moderate amounts of external corrosion.

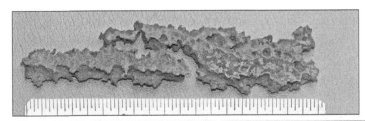

FIGURE 4.8
(See Colour Insert.) Loose sample acquired during inspection: Views of the internal surface where the pitting attack had initiated are shown. The surfaces formed by the corrosion attack are highly irregular and contain darkish deposits.

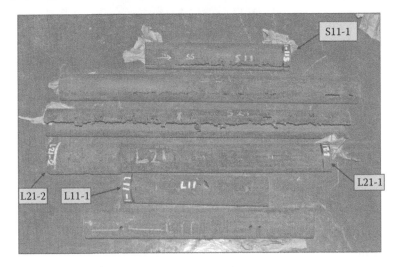

FIGURE 4.9
Samples delivered: Representative specimens extracted for examination/chemical analysis are S11-1 from an SS tube, and L11-1, L21-1, and L21-2 from LS tubes; L21-2 was a collar specimen.

4.2.7 Hardness Tests

Hardness tests were conducted on a laboratory Rockwell machine. The results are presented in Table 4.4.

The results conform to API 5CT specifications, which require the maximum hardness to be 23 HRC.

(a)

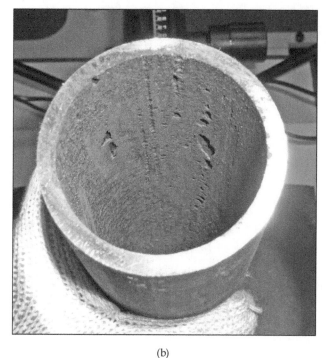

(b)

FIGURE 4.10
(See Colour Insert.) (a) A half-section from sample L11. (b) The internal surface of L11 with well-defined pits containing darkish deposits.

4.2.8 Metallographic Examination

Specimens S11-1, L11-1, and L21-2 were subjected to optical microscopy after the required preparation. Firstly, the basic microstructure of the specimen was evaluated, then the nature of damage and surface scale.

Specimen S11-1:

- The microstructure of the parent metal was a fine martensite without any abnormalities (Figure 4.12a,b).

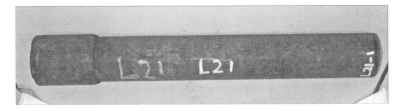

(a)

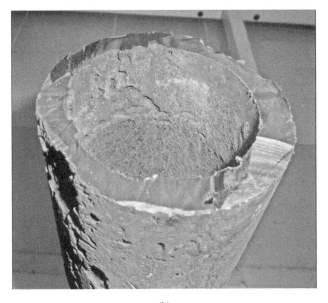

(b)

FIGURE 4.11
(See Colour Insert.) (a) Sample L21 after cutting off specimen L21-2 from the collar. (b) Sample L21: Examination shows that the internal surface of the collar is also badly pitted and covered with darkish deposits. The bright debris on the deposits are metal particles that were formed from the sawing process.

- The darkish scale on the surfaces of the internal pits showed a distinct crystal structure under crossed Polaroids (Figure 4.12c,d). This shows that the scale was not composed of the normal oxidation products of steel.

Specimen L11-1:

- The parent metal microstructure was similar to that of S11-1 and is not illustrated.
- The scale on the internal pits also had a crystal structure, but was mixed with some less crystalline forms (Figure 4.13a,b).

TABLE 4.1

Laboratory Testing Program

Specimen	XRD	Gas Chromatography	Chemical Analysis	Hardness and Microstructure	SEM/EDS
S11-1	Yes	Yes	Yes	Yes	Yes
L11-1	Yes	No	Yes	Yes	Yes
L21-1	No	No	Yes	No	No
L21-2 (collar)	No	No	Yes	Yes	Yes

TABLE 4.2

Gas Chromatography Analysis on Sample S11

Gas Test	Result (ppm v/v)
Hydrogen (H_2)	379,201
Oxygen (O_2)	60,245
Nitrogen (N_2)	217,488
Methane (CH_4)	673
Carbon Monoxide (CO)	18
Carbon Dioxide (CO_2)	947,710
Ethylene (C_2H_2)	85
Ethane (C_2H_6)	103
Acetylene (C_2H_2)	151

TABLE 4.3

Chemical Analysis of API 5CT Samples of LS and SS Tubes

Element	C	Mn	P	S	Si	Ni	Cr	Mo	V	Al	Cu
L11	0.23	1.31	0.015	0.015	0.27	0.021	0.167	0.002	0.002	0.039	0.013
L21	0.23	1.31	0.015	0.019	0.263	0.021	0.167	0.002	0.002	0.035	0.013
S11	0.23	1.31	0.021	0.014	0.266	0.020	0.199	0.03	0.000	0.030	0.012

TABLE 4.4

Hardness Results

	Rockwell Hardness (HRC)			
Tube Sample	1	2	3	Average
L21	16.64	17.60	17.70	17.3
L11	17.91	17.80	17.80	17.8
SS11	17.55	18.60	18.73	18.3

(a)

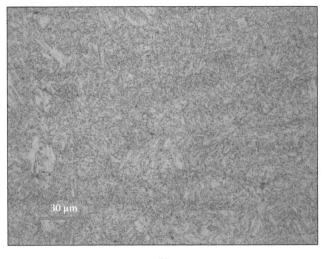

(b)

FIGURE 4.12
(a) S11-1: Microstructure of material is difficult to resolve at this magnification. Mag. 200×.
(b) S11-1: Magnified view shows the microstructure to be fine-tempered martensite. Mag. 500×.
(c) S11-1, internal pit: Section across heavy pitting; deposits have a crystal structure. Mag. 100×.
(d) S11-1, internal pit: Magnified view of crystal structure. Mag. 200×.

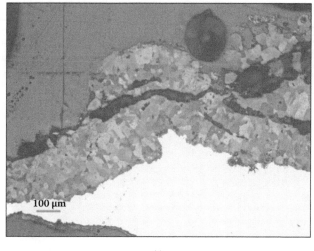

(c)

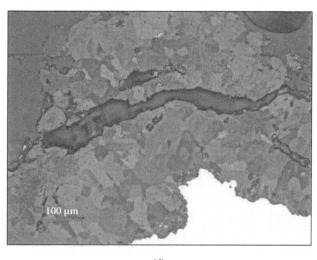

(d)

FIGURE 4.12 (Continued)

Specimen L21-1:

- Similar to L11-1, the parent metal microstructure was similar to that of S11-1 and is not illustrated.
- The scale on the pits was similar to S11-1 (Figure 4.14a,b).

4.2.9 SEM/EDS Analysis

The specimens listed in Table 4.1 were examined under a scanning electron microscope and the deposits were subjected to elemental analysis using the

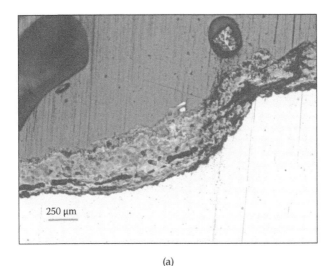

(a)

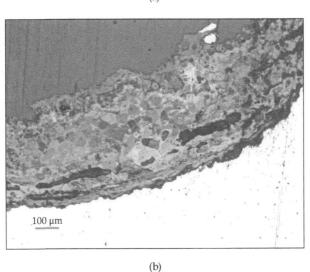

(b)

FIGURE 4.13

(a) Specimen L11-1, internal pit: Section through a pit and deposits. Mag. 50×. (b) Specimen L11-1, internal pit: Magnified view shows that the deposits have a crystal structure mixed with some less crystalline forms. Mag. 100×.

built-in energy-dispersive X-ray spectrometer (EDS). An examination was performed on the surface scale as well as on their cross-sections (metallographic specimens).

The results are presented in Table 4.5. It can be seen, that in all cases the dominant elements detected on the dark scale were Fe, C, and O; on the cross-sections where contamination was minimal, they were the only species detected.

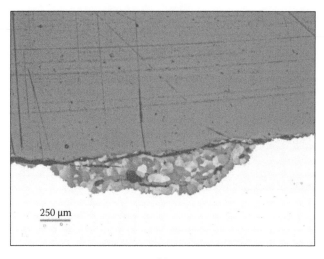

(a)

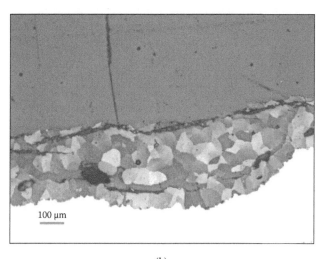

(b)

FIGURE 4.14
(a) Specimen L21-1, collar: Section through an external pit and deposits. Mag. 50×. (b) Specimen L21-1, collar: Magnified view shows that the deposits have a crystal structure. Mag. 100×.

4.2.10 Discussion

It was shown that both the SS and LS tubes and the affected collars had experienced pitting on their internal surfaces. In all cases, the pits contained a darkish scale.

Despite the presence of pits on the external surfaces of the LS tubes and the collars, it is very likely that all the pits had originated from the internal

TABLE 4.5

Composition of Element in Specimens
S11, L11, and L21 (from SEM/EDS)

Specimen	Element	Composition (%wt)
S11	C	24.40
	O	30.59
	Fe	45.01
L11	C	23.62
	O	25.86
	Fe	50.52
L21	C	32.92
	O	29.47
	Fe	37.60

surfaces, and that the external pitting had been caused by products that had leaked out from the tubes after the tube wall had been breached.

The chemical composition, hardness, and microstructure of the specimens tested showed that the tubes conformed to API 5CT specifications.

Optical microscopy showed that the darkish scale had a distinct crystal structure, not shown by normal corrosion products of steel.

Gas chromatography showed that the scales (brown and dark combined) gave off large amounts of CO_2 on reaction with HCl; $FeCO_3$ scale would give off CO_2 but corrosion oxides would not.

EDS of the dark scale showed that Fe, C, and O were the dominant elements; in fact, on cross-sections of the scale where contamination was minimal, they were the only elements detected.

It was also deduced that the dark scale was $FeCO_3$. This product would have arisen from CO_2 corrosion, where the presence of CO_2 and water would form acidic H_2CO_3 (carbonic acid) and attack the steel. Such corrosion is common in oil and gas components. More information can be referred to in API 571 [3].

The pattern of attack, where the pitted areas were distinctly separated from non-affected areas, has often been called *mesa* attack [4]. The attack pattern is caused by local breaching of a protective $FeCO_3$ scale, which then precipitates pitting at the breached locations. The breaching could be caused by excessive turbulence or by abrasives caught in the flow stream.

In this case, it was observed that the damage was only confined to one side of the tubes, and that only the inclined parts of the tubes were affected. This suggests that abrasives (corundum) flowing over the bottom of the tubes at the incline were responsible.

4.2.11 Conclusion

- The corrosion damage on the SS and LS tubes and collars were all due to the CO_2 attack, of a form known as a *mesa* attack. A CO_2 attack requires the presence of sufficient amounts of CO_2 and water in contact with the surface of the steel.

- It is likely that that the attack, which occurred only at the inclined parts of the tubes, had been precipitated by the flow of entrained corundum.
- The materials used conformed fully to specifications.

4.3 Case Study 2: Tin Dredge Wheel Pinion Failure

4.3.1 Background Information

The pinion wheel is part of the gearing system that drives the bucket band in a tin dredge—in the late 1970s, two teeth had broken off from such a pinion during operation. Failure was typically fatigue in nature and had initiated from the fillets at the roots of the teeth. A site examination of the pinion showed that fatigue cracks were present at the roots of many other teeth (see Figure 4.15). A similar pinion used on the other side of the bucket drive was cracked in the same way.

From the available information, the pinions were put in about five years earlier and were driven by a 600 hp motor (note that 1 hp = 0.746 kW). The motor was replaced by a larger one with 1000 hp. The pinion failure occurred after about six to seven months of operating time with the increased power.

FIGURE 4.15
The presence of fatigue cracks on other teeth. Note also the darkish heat marks due to attempted flame hardening.

4.3.2 Physical Examination

A broken tooth specimen was taken to the laboratory for examination. This tooth was broken into two pieces of about 10 inches and 6 inches in length. Relevant features are as follows:

- The fracture surfaces showed a typical fatigue fracture. Fatigue cracks had initiated from both sides of the tooth (see Figures 4.16 through 4.18) and extended inwards. The cracks were of unequal length. The crack at the compression side was shorter and showed sign of greater age (more pronounced surface smoothening). Both cracks had multiple origins.

- Surface wear was present on both sides of the tooth. However, the wear was unequal; the compression side was only slightly worn (see Figure 4.19) whereas the tension side was heavily worn, especially at the pitch line (see Figure 4.20). Pitting was also evident on this surface.

FIGURE 4.16
These images show the general location of cracking. Note also the darkish heat marks.

FIGURE 4.17
The fracture surface of one end of the tooth. Note the unequal length of fatigue cracks. Note also the multiple crack origins.

FIGURE 4.18
The fracture surface of the other end of the tooth. Note the welding repair at the extreme end.

FIGURE 4.19
The side of the tooth with less wear (compression side). Note the non-uniform wear and deformation at tip of the tooth.

FIGURE 4.20
The opposite side of the tooth with heavy wear (tension side). Note the presence of pitting.

- The wear on the face was also quite uneven; one end of the tooth was worn much more than the other (compare Figures 4.19 and 4.21). This shows the presence of a gross misalignment.
- A deformed band was evident at the top surface of each side of the tooth. The deformation had caused the material to flow and form

FIGURE 4.21
The same side as in Figure 4.19 but at the other end. Note the absence of wear. The tip, however, is deformed, but less than that shown in Figure 4.19.

FIGURE 4.22
Deformation at the top part of the surfaces of the tooth.

two small projections at the tip of the tooth (see Figure 4.22). This had resulted from an inaccurate tooth profile.

- One end of the tooth (see Figure 4.18) contained casting defects which had been repair welded and roughly ground to shape. This repair was not completely successful but was not a main factor in initiating the fatigue cracks.

- Darkish heat marks present on the surfaces of the tooth showed that an attempt had been made to surface-harden the teeth with a flame (see Figures 4.15 and 4.16).

FIGURE 4.23
Fine cracks at the root as a result of a poor machining finish.

- The tooth was finished by planing. The cutting process was rough and induced fine cracks at the roots perpendicular to the planing direction (see Figure 4.23). The pattern of the fatigue crack growth showed that these initial machining cracks helped to initiate the fatigue.

4.3.3 Microscopic Examination

Various sections were taken and examined with the following results:

- The material at the core had a medium-grained ferritic/pearlitic microstructure (see Figure 4.24) with a carbon content of about

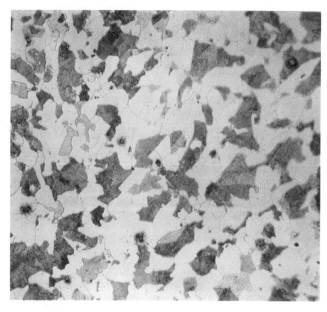

FIGURE 4.24
Microstructure of the core material consisting of ferrite and pearlite.

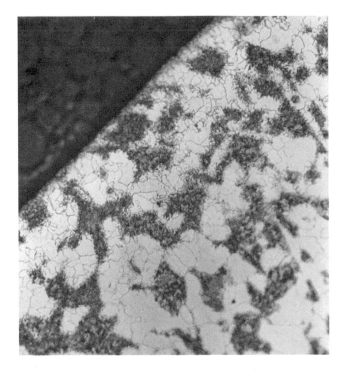

FIGURE 4.25
Spheroidization of pearlite at the 'hardened' surface layer as a result of insufficient heat. Note the martensitic patches within certain pearlite grains.

0.30/0.35% and an average hardness of about 163 HB. It was in an annealed condition. This steel conforms substantially to BS 592 Grade C specifications.

- The material quality was acceptable.
- The 'hardened' surface also consisted of ferrite and pearlite. Breakdown of the pearlite grains had started to occur; this had been brought about by heating to just below the transformation point. Isolated patches within certain grains had however transformed to martensite (see Figure 4.25).
- The surfaces at the pitch line where heavy wear had occurred were deformed showing that the local contact stresses had exceeded the yield strength of the steel (see Figure 4.26).

4.3.4 Stress Analysis

The bending and surface contact stresses were examined as there were clear signs of overloading. Two methods of stress calculations were used; firstly a British method according to BS 436 (1940) [5], and secondly an American method according to AGMA (American Gear Manufacturers Association) [6].

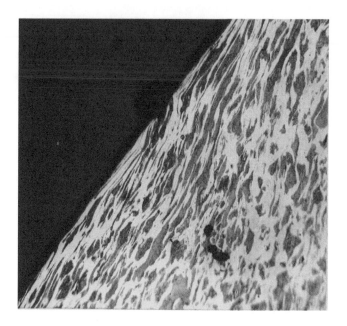

FIGURE 4.26
Deformed grains at wear surfaces of the tooth due to excessive contact stress.

Details of the calculations are provided in Appendices 4.1, 4.2, and 4.3. The results are presented in Tables 4.6 and 4.7.

An examination of the results summarized in the two tables shows the following:

- Both methods predicted that BS 592 Gr. C was unsafe in wear, but whereas the BS 436 method indicated a surface-hardened BS 970:1955, EN9 steel to be suitable, the AGMA method required a case carburized steel.

- BS 436 considered the BS 592 Gr. C steel to be suitable for strength whilst AGMA considered it highly unsuitable. AGMA requires a case carburized steel for safety. The 'strength' considered is the fatigue strength.

4.3.5 Discussion and Conclusion

1. Observed results
   - Failure was due to fatigue cracks which had initiated at the fillets at both sides of the tooth. The crack at the compression side was shorter and older, and had probably formed when the pinion was used in the reversed position originally. Considering the presence of similar cracks in other teeth it may be deduced that

TABLE 4.6

Results of BS 436

Criterion of Design	Results of Calculations	Material Choice
a. For wear		
$S_c \geq \dfrac{KF_t}{bX_cZ}$	$\dfrac{KF_t}{bX_cZ} = 4004$	A material with $S_c \geq 4004$ is required. En9, surface-hardened, will give $S_c = 4000$.
Where S_c is the allowable surface stress factor for the material used. For BS 592 Gr. C, $S_c = 1400$.	$Sc = 1400 < 4004$ BS 592 Gr. C is unsafe in wear	
b. For strength		
$S_b \geq \dfrac{PF_t}{bX_bY}$	$\dfrac{PF_t}{bX_bY} = 12,919$	BS 592 Gr. C is suitable.
Where S_b is the allowable bending stress factor for the material used. For BS 592 Gr. C, $S_b = 19,000$.	$S_b = 19,000 > 12,919$ BS 592 Gr. C is safe in strength	

TABLE 4.7

Results According to AGMA

Criterion of Design	Results of Calculations	Material Choice
a. For wear		
$\sigma_c \geq S_{ac}\left(\dfrac{C_L C_H}{C_T C_R}\right)$	$S_{ac}\left(\dfrac{C_L C_H}{C_T C_R}\right) = 62,400$	A material with $S_c \geq 227,063$ is required. This can only be achieved with a carburized case with $H_v \geq 700$ (or HRC ≥ 60).
Where σ_c = calculated contact stress number given by, $\sigma_c \geq C_p\left(\dfrac{F_t C_o C_s C_m C_f}{C_v dbI}\right)^{1/2}$	$\sigma_c = 181,651 > 62,400$ Material (BS 592 Gr. C) is unsafe in wear	
b. For strength		
$\sigma_t \leq S_{at}\left(\dfrac{K_L}{K_T K_R}\right)$	$S_{at}\left(\dfrac{K_L}{K_T K_R}\right) = 12,667$	A material with 70,806 psi is required. This can only be achieved with a carburized case with $H_v \geq 700$ (or HRC ≥ 60).
Where σ_t = calculated stress at root, given by, $\sigma_t \geq \dfrac{F_t K_o PK_s K_m}{K_v bJ}$	$\sigma_t = 47,204 > 12,667$ BS 592 Gr. C is safe in strength	

the operating stresses had exceeded the fatigue strength of the steel in bending by a large amount.

- A large amount of misalignment was present, as shown by the uneven wear and deformation patterns. Further, the nature of deformation at the sides of the tooth tip showed that the tooth profile was originally inaccurate. The presence of deformation at the pitch line is also proof that the surface contact stresses had exceeded the compressive yield strength of the material.

- The steel conformed substantially to BS 592 Grade C, and was of acceptable quality. However, some large casting defects were present at the end of the tooth which had been repair welded and then roughly ground down, both unsatisfactorily. The core material was in annealed condition. An attempt had been made to harden the surfaces of the teeth but it was unsuccessful. Only small patches within certain pearlite grains had transformed to martensite whereas the remainder of the material was slightly softened. A further negative aspect was that a residual tensile stress system would have to be set up at the surface, which would add to the operating stress.

- The surface finish at the roots was bad; rough machining marks and fine cracks were present which had lowered the fatigue strength of the steel and contributed towards the initiation of the cracks.

2. Comparison of BS 436 and AGMA methods of stress calculations

- Surface durability: Both methods predicted correctly that BS 592 Gr. C would not be sufficient to resist in the pinion under study. However, there are significant differences between the BS and AGMA recommendations on what is required; the BS method considers a surface-hardened EN9 steel to be sufficient whereas the AGMA method requires a case carburized steel. Considering the shock loading and vibration present in the operation of a dredge, together with the presence of misalignment between the pinion and gear, and the inaccuracy and bad finish of the teeth, the AGMA method appears to be more representative. It at least attempts to incorporate these factors into the equations whereas the BS method does not make any allowances for them at all.

- Fatigue strength in bending: It is quite clear that the fatigue strength of the steel had been grossly exceeded, yet this is only predicted by the AGMA equation. The BS method considers the pinion to be safe by quite a large margin. There is no doubt that the AGMA method is the more accurate, for the same reasons given in the methods comparison in surface durability. The fact that the BS method predicts safety for the 1000 hp operation whereas fatigue cracks had already developed with the 600 hp operation is an indication of how unsuitable this method is. The AGMA method again requires a case carburized steel for safe operation.

3. Comments

Based on the AGMA method in estimating the safety from wear and fatigue, carburized steel with a case hardness of 700 HV (60 HRC) should be used. However, this is impracticable because the local

industry is unable to manufacture such a gear to the required toler-
ance and finish required and to case harden and heat-treat it successfully. Therefore, alternatives to this have to be found.

It should not be assumed that the BS standards are inferior in any way to
others, as the BS 436 edition used was published in 1940, when gear design
technology was still wanting. Unfortunately, the local dredging industry was
still using it throughout the 1970s, until alerted by the Failure Investigator
that other design standards were more appropriate. BS 436:1940 was replaced
in 1986 by BS 436-3:1986, which was more representative of the state of the
technology at that time.

4.3.6 Prevention against Fatigue

The basic problem is that various factors, which have been discussed, magnify operating stresses and at the same time reduce the fatigue strength
of the tooth. Therefore, if these factors can be alleviated, a lower strength
material can be used. This may be achieved by the following methods:

- Annealing the gear blank before machining, to remove any residual
  casting stresses as well as to homogenize the material.
- Cutting the tooth profile accurately.
- Ensuring that the surface finish at the fillets is free of rough machin-
  ing marks and cracks.
- Performing an appropriate surface-hardening process.
- Performing a final profile grinding to correct any distortions result-
  ing from surface-hardening.
- Increasing the fillet radii. It has been shown that an increase from
  $\frac{0.2}{P}$ to $\frac{0.4}{P}$ can increase the fatigue strength of hardened gears by
  20% (where P = diametral pitch).
- Aligning the pinion and gear accurately and having firm rigid supports.

It must be taken into account that most of these measures cannot be
accurately quantified but it is expected that a net reduction in the effective
stresses of about 40% can be achieved. A low-alloy surface-hardening steel
can then be used. A suitable cast steel is the German GS-50 Cr Mo 4 (no com-
parable BS) with the following specifications:

- Chemical composition (%): 0.46–0.54 C, 0.25–0.50 Si, 0.50–0.80 Mn,
  0.90–1.2 Cr, 0.15–0.25 Mo
- Tensile strength (40 mm cast on bar): 47 tsi or equivalent to 648.11
  MPa (minimum)

- Surface hardness: 50 HRC (minimum) (about HV 550)
- Depth of hardness penetration: 5 mm (minimum)
- Heat treatment: Quenched and tempered to a hardness level of between 240 to 260 HV at the core
- Weldability: Special precautions must be taken

4.3.7 Prevention against Surface Wear

Performing the measures outlined in Section 4.3.6 will also reduce the surface contact stresses to some extent. The steel recommended should be suitable for wear.

4.3.8 Precautions

Certain important precautions must be observed or success cannot be ensured. These are as follows:

- Surface-hardening must extend over the fillets and roots of the teeth otherwise fatigue will certainly occur. The quality of hardening is critical and must be closely checked.
- Castings with defects present near the roots and fillets must be rejected for two reasons: firstly, repair welds are seldom 100% satisfactory; secondly, even if the welding was perfect, the weld metal would still be much softer than required and fatigue cracking would eventually occur.
- The fillet and root surfaces must be 100% checked for cracks and small flaws before and after hardening.

Furthermore, it is advisable to check regularly for cracking at the tooth fillets, preferably with the aid of a penetrant dye or a magnetic particle crack detector.

4.4 Case Study 3: Excessive Pin Wear in Tin Mining Dredge

4.4.1 Background Information

In a tin mining dredge, an endless bucket band is used to dig up the mineral deposits at the bottom of a large, man-made pond. The buckets are linked together by pinning; the pins are fixed at either end to the two spaced front eyes of the bucket. The barrel of the pin is inserted within the back eye of the adjacent bucket, which is lined with an austenitic manganese steel (AMS) bush. During operation, the pin rotates within the stationary bush and the contacting parts of both the pin and bush are subjected to heavy wear. The pin is machined from a high-strength bainitic/martensitic steel (BS 970: 1955, EN 24 or 26), from forged stock or a casting.

A set of 15 new pins were found to have suffered from excessive wear. These pins had only run for about 2432 hours but had suffered from wear of about 38 mm which is more than double normal expectations. Old, rewelded pins in the same band exhibited normal wear.

The new pins had been bought from the same source as the previous ones but the bushes were from a different supplier.

Two of the worn new pins were examined at the workshop. Except for the excessive wear, the pins appeared to be normal. The diameter of the pin was about 202 mm and the length of the barrel was about 889 mm long. The manufacturer's test certificate showed that the chemical composition, heat treatment, and hardness of the pins were within specifications. One of the worn pins (Pin No. 1) was later sectioned for macro- and micro-examination.

A worn bucket bush was also examined. This bush was about 11 mm thick at the thinnest part of its edges and still had some life left. One edge was badly cracked but otherwise no abnormalities were evident. Samples were taken from the bush for metallographic examination and hardness testing.

A worn old pin (Pin No. 2), which had exhibited the normal wear pattern was also provided for examination. This pin was expected to serve as a reference.

According to the mine manager, the new pins were subjected to the same operating conditions as the old ones. In particular, stripping speeds varied between 28 and 35 buckets per minute and normal operating speeds, between 18 to 24 buckets per minute. Normal intervals for pin and bush changes varied between 2400 and 2500 hours. Previous new pins only wore down between about 7 and 13 mm in this period, whilst rewelded pins wore down about 7 mm.

At a later date, two new pins which had run for 1192.5 hours were supplied for visual examination. One pin had mated with the bush No. 1 and the other with the bush No. 2. The pin with the bush No. 1 had a maximum wear of 4 mm whereas the other one suffered from a maximum wear of 6 mm and in addition showed more surface damage. A linear extrapolation will give a pin wear of 8.2 mm with the A bush and 12.2 mm with the bush No. 2, in 2432 hours which is within normal limits.

4.4.2 Specimens for Performing Hardness Tests and Micro-Examination

The following sections from Pins No. 1 and No. 2 were taken for examination.

- Plane A, situated about 22 mm from the end of the pin away from the head.
- Plane B, about 125 mm away from Plane A.
- Plane C, at the mid-point of the barrel of the pin.

The sections to be examined were all finely machined and finished with fine emery paper giving conditions suitable for hardness testing, sulphur printing, and micro-etching.

4.4.3 Hardness Test for the Pins and Bushes

Hardness tests were performed on all three sections from each pin. Tests were performed on a Brinell machine at a load of 3000 kg using a 10 mm diameter ball. Certain areas were rechecked using a Vickers machine at a load of 30 kg.

Hardness tests were first taken around the circumference at a depth of about 8 mm from the surface of the pin barrel. Then, a hardness traverse was taken along the diameter joining the mid-points of the tension and compression (wearing) surfaces of each section. Indentations were spaced about 12 mm apart.

Details of hardness values are given in Table 4.8. It may be seen that hardness values in both pins vary with position. The areas which were harder were subjected to better quenching during heat treatment. Pin No. 2 was generally harder than Pin No. 1 and the ends of both pins were generally harder than the mid-points. In Pin No. 1, on the compression side at Plane C (mid-point of pin), the hardness was constant at about 311 HB from the surface to a depth of 40 mm. In Pin No. 2, the same pattern prevailed but the hardness was higher at 320 HB.

TABLE 4.8

Hardness Traverses (HB 3000)

Distance from Tension Surface (mm)	Pin No. 1			Pin No. 2		
	Plane A	Plane B	Plane C	Plane A	Plane B	Plane C
8	320	311	311	340	330	320
20	311	311	311	340	330	320
32	311	303	302	339	330	320
44	311	303	302	339	330	320
56	302	304	302	339	330	320
68	302	302	302	352	340	330
80	302	302	302	352	352	339
92	293	293	293	358	352	340
104	302	286	285	360	352	340
116	302	304	293	358	340	328
128	311	304	302	358	330	313
140	311	305	307	352	330	320
152	311	309	305	349	330	320
164	311	307	311	347	330	320
176	311	309	—	352	330	320
188	320	311	—	352	339	—

For the bush, Vickers hardness values were made on prepared micro-specimens. Values at non-deformed areas ranged between 250 and 280 HV (10 kg) with an average value of 264 HV (10 kg). The corresponding average Brinell value would be about 260 HB.

4.4.4 Macro-Examination and Sulphur Printing

An examination of the worn surfaces at the mid-points of the pins (Plane C) showed that the surface of Pin No. 1 contained grooves and cavities which were both larger and more numerous than in Pin No. 2. The wear surface on Pin No. 2 contained a larger smooth area.

The worn surfaces of the bush were free of grooves and cavities and were relatively smooth.

Plane B of each pin was subjected to sulphur printing. The amount and distribution of sulphur in both pins was fairly similar.

All three planes of both pins were macro-etched and examined. No abnormalities were detected in any of them and no differences were evident between them.

4.4.5 Micro-Examination of Pins No. 1 and 2

Micro-specimens were taken at various locations on Plane C of both pins. The following general observations may be made:

- The microstructure of both pins basically consisted of tempered bainite with small amounts of ferrite (see Figure 4.27 through Figure 4.30). Small amounts of manganese sulphide inclusions of various shapes were also present.

- At the centre section of Pin No. 2, harder and more acicular forms of bainite appeared (see Figure 4.31). However, for Pin No. 1, softer, coarse, tempered bainite was present, together with ferrite (see Figure 4.32). These differences were deemed to have arisen from differences in the quenching mechanics; Pin No. 1 had a poorer quench than Pin No. 2 during heat treatment.

The wear surfaces of both pins showed fairly similar features. With reference to Figure 4.33 through Figure 4.35, these may be detailed as follows:

- A surface layer about 0.6 mm deep had undergone intense plastic deformation by shearing. At certain areas, parts of this layer had transformed into a featureless, white etching microstructure which was probably newly formed martensite. This showed that the shearing had been so intense that the heat generated in the layer had

FIGURE 4.27
Pin No. 1, at a depth of 5 mm below the tension surface. The microstructure comprises tempered bainite with small amounts of ferrite. Mag. 400×.

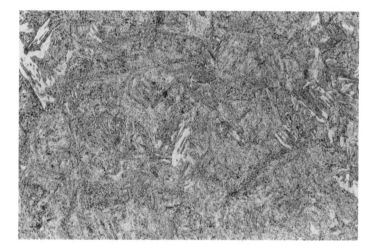

FIGURE 4.28
Pin No. 2, at a depth of 5 mm below the tension surface. The microstructure comprises tempered bainite with small amounts of ferrite. Mag. 400×.

raised its temperature to beyond 700°C. Manganese sulphide inclusions in the deformed layer had correspondingly been flattened into sharp-edged, crack-like planes.

- Cracks were observed to have formed in the deformed layer and assisted in wear particle formation. Crack formation appeared to have been helped by the weak manganese sulphide planes which essentially behave as cracks.

FIGURE 4.29
Pin No. 1, at a depth of 40 mm below the compression surface. The microstructure comprises tempered bainite with small amounts of ferrite. Mag. 400×.

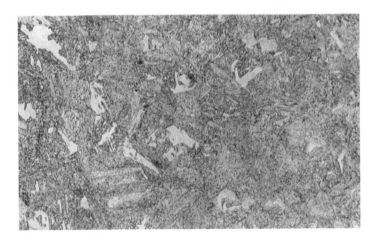

FIGURE 4.30
Pin No. 2, at a depth of 40 mm below the compression surface. The microstructure comprises tempered bainite with small amounts of ferrite. Mag. 400×.

- The maximum hardness in the deformed layer occurred at the white martensite where values as high as 670 HV (200 g) were measured. Other areas had maximum hardness values of only about 575 HV (200 g).

4.4.6 Micro-Examination of the Bucket Bush

Micro-specimens were taken at various locations in the bush. The microstructure consisted of austenite grains with some grain boundary carbides

FIGURE 4.31
Pin No. 2, at the centre. The microstructure comprises acicular tempered bainite with small amounts of ferrite. Mag. 400×.

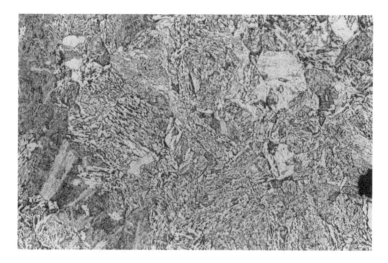

FIGURE 4.32
Pin No. 2, at the centre. The microstructure comprises coarse tempered bainite with small amounts of ferrite. Mag. 400×.

and globular carbides (see Figure 4.36). This is a fairly common microstructure in bucket bushes.

The wear surface also contained a layer of intense plastic deformation. This layer had a maximum hardness of about 730 HV (200 g).

Large amounts of shrinkage defects were present which had contributed to the easy cracking of the bush.

FIGURE 4.33
Pin No. 1, worn surface. Showing the deformed layer and crack information. Mag. 50×.

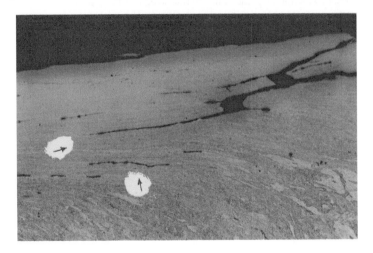

FIGURE 4.34
Magnified view of the left part of the crack from Figure 4.33. The arrows point to flattened manganese sulphide inclusions. Mag. 200×.

4.4.7 Discussion

1. Old pin: This pin (Pin No. 1) is expected to serve as a reference. Chemical composition and heat treatment details were not available but the microstructure showed that it was in normal condition. Two basic points may be made. Firstly, the microstructure was fairly uniform throughout the section, though at the centre, harder acicular bainitic forms were present; and secondly, hardness values to a

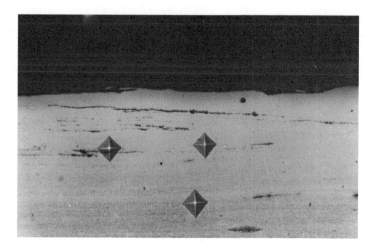

FIGURE 4.35
A magnifed view of the top, centre-left of Figure 4.33, showing micro-hardness indentations, flattened manganese sulphide planes, and a white structure. Mag. 400×.

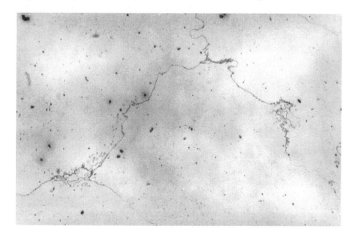

FIGURE 4.36
Microstructure of the bush material consisting of grain boundary carbides and undissolved carbides in an austenitic matrix. Mag. 400×.

depth of 20 mm ranged between 320 and 352 HB with the harder values occurring at the end of the pin. The specified hardness values for this pin are not known, but its higher hardness range exceeds the current specification of 330 HB maximum.

2. New pins: The chemical composition and heat treatment of the new pins conformed to specifications. The general microstructures of the samples examined were similar to that of Pin No. 2 to a depth of at least 40 mm. The sulphur prints and micro-sections were similar

to those of Pin No. 2. The hardness readings to a depth of at least 20 mm from the surface of the pin barrel ranged between 311 and 320 HB, again with the higher values occurring at the end of the pin. These values are within specifications and should have been quite adequate based on past experience. Deeper into the section however, more ferrite and a coarser bainite were present. These are softer phases and the hardness levels correspondingly dropped to minimum values of about 285 HB at the centre of the section. However, this finding is of little consequence to the wear problem because the pin is not expected to wear so deep.

3. Bush material: This material was austenitic manganese steel with a fairly normal microstructure. The average hardness of the undeformed material was about 260 HB which is about 10% higher than usual. However, this is still acceptable for austenitic manganese steel.

4. Experimental studies of adhesive wear mechanisms: Laboratory studies have shown that in unlubricated or partially lubricated sliding between metals, wear is a function of load. In many metals, the 'mild wear' which occurs at low speeds and loads may suddenly transform to 'severe wear' when speeds or loads, or both, are gradually increased. In mild wear, the surface is typically smooth and undamaged, whereas in severe wear, the surface experiences heavy tearing and damage. Wear rates at the severe wear regime are typically two orders of magnitude (i.e., 100 times) higher than at the mild wear regime. The transition from mild to severe wear also depends upon the operating environment and the hardness of the metal. Environments which promote strong welded junctions favour early transitions and vice versa. More information on adhesive wear can be found in ASM Vol. 18 [7].

Increasing initial metal hardness will delay the onset of transition and reduce the level of severe wear. At a high enough hardness, the transition to severe wear may even be prevented, though hardnesses in excess of 350 HV would induce unacceptable brittleness. In some instances, mixed wear can occur where both mild and severe wear mechanism operate. Small changes in environment, operating conditions, or hardnesses may induce one type of wear or the other, to predominate with the attendant effects on the wear rate.

5. Observed wear mechanisms in the pins: The wear surfaces of both pins showed the presence of a mild and severe wear mechanism, though in Pin No. 1 the severe form was much more predominant. The mode of wear particle formation was however similar in both pins; intense plastic deformation and work hardening occurred at a layer on the wear surface, which developed micro-cracks, eventually extending to form wear particles. Mircro-cracking had been assisted

by the presence of weak manganese sulphide planes but the degree was approximately similar in both pins. The maximum hardnesses in the deformed layers of both pins were about equal, at about 670 HV (200 g). These observations indicate that the excessive wear of Pin No. 1 had not been primarily due to its hardness or microstructure.

6. Observed wear mechanisms in the bush: This appeared to be quite similar to that found in the pins, except that the manganese sulphides were not significant and that the maximum hardness of the deformed layer was about 730 HV (200 g). This is normal for austenitic manganese steel bushes.

   From the above observations, together with the finding that Bush No. 2 also suffered from more wear than Bush No. 1, showed that its slightly higher initial hardness had not been a major factor.

7. Possible reasons for excessive wear in Pin No. 1: This study is too limited to allow the exact causes to be positively identified. However, it is clear that the severe wear form had been more predominant in Pin No. 1. The following factors are thought to have contributed:

   • Slightly lower than usual pin hardness.
   • Slightly higher than usual bush hardness.
   • Unknown changes in the environment or the chemical composition of Bush No. 2, which encouraged the formation of stronger adhesion between the pin and bush.
   • Unknown changes in operation.

   From the observed wear mechanism in the pins and bushes, and also from later results which showed that two new pins mated to Bushes No. 1 and No. 2 wore down 4 mm and 6 mm, respectively in 1192.5 hours, the above first two factors can be considered to have only minor effects and could not have been solely responsible for the excessive wear.

   Thus, it appears that either the third or fourth factor above or a combination of both had been the main causative factor. Furthermore, clarification is not possible at this stage.

8. Remedial action: The batch of heavily worn down pins can be reused after welding build up. It is suggested that two buttering layers of a suitable stainless steel be first deposited.

If the causes of the excessive wear need to be known, more research is needed. This would involve close monitoring of operating conditions, environment, wear rates, and further metallographic studies, especially on the different bushes.

4.4.8 Conclusion

- The hardness and microstructure of the new pins are within specifications.
- The hardness values of the new pins and Bush No. 2 have not been a major factor in causing the excessive wear.
- The causative factors are suspected to be either the third factor or fourth factor in item 7 of the 'Discussion' Section 4.4.7, 'Possible reasons for excessive wear in Pin No. 1' or a combination of both. However, further clarification is only possible after further investigation.
- The 14 remaining excessively worn pins can be reclaimed by welding but care must be exercised.

Problems and Answers

Problem 4.1

The data for a failed wheel pinion from the tin dredge are given as in the following:

Spur gear, 20° pressure angle

No. of teeth (T)	= 25, mating on 152 teeth
Speed (N)	= 28.73 rpm
Pitch diameter (d)	= 27.852
Face width (b)	= 16″
Motor power	= 1000 hp
Material	= BS 592 Grade C
Transmitted tangential load (F_t)	= 35.16 T = 78,758 lb

Diametral pitch (P) $= \dfrac{25}{27.852} = 0.8976 \text{ tooth/in diameter}$

Pitch line velocity (v_P) $= \pi Nd = 209.49 \text{ ft/min}$

Calculate the pinion wear surface durability and fatigue strength by using the BS 436 and AGMA standards and give comments from the calculated values.

Answer 4.1

Refer to Appendices 4.2 and 4.3.

Problem 4.2

A number of bucket bushes in a tin dredge move slowly and operate under a high load. If you are required to choose a surface rubbing material: between lead, polytetrafluoroethylene (PTFE), and polyethylene for the pin to be fitted to the steel bucket bush, which of the following materials will give the longest wear life? Given that, the hardness for lead, PTFE, and polyethylene are 30 MPa, 50 MPa, and 70 MPa, respectively. Additionally, the adhesive wear coefficient for the rubbing material: lead, PTFE, and polyethylene on steel are 2×10^{-5}, 2×10^{-5} and 1×10^{-7}, respectively. Comment on the suitability of your results for practical application.

Answer 4.2

The wear volume is, $v = \dfrac{kWL}{3H}$

where, k is the adhesive wear coefficient of a material on steel, H is the hardness, L is the sliding distance, and W is the normal applied load.

The wear volume is minimum when $\dfrac{H}{k}$ is maximum. Therefore,

For lead, $\dfrac{H}{k} = \dfrac{30(10^6)}{2(10^{-5})} = 1500 GPa$

For PTFE, $\dfrac{H}{k} = \dfrac{50(10^6)}{2(10^{-5})} = 2500 GPa$

For polyethylene, $\dfrac{H}{k} = \dfrac{70(10^6)}{1(10^{-7})} = 700,000 GPa$

From the calculations, it is obvious that polyethylene should give the longest life wear compared to PTFE and lead.

This example is included to illustrate the dangers of a theoretical study without knowing the actual operating conditions. The use of polymers has been attempted before by the very enterprising engineers at the then Malayan mining companies in the early 1970s, and all the experiments failed miserably. The problem was that the bearing stresses were calculated by just dividing the maximum tension of the bucket band by the projected area of the bushes. This method did not consider that the pin was relatively long and that the elastic bending caused during operation resulted in much higher localized stresses at the edges of the bush material, which exceeded its flow stress and caused it to squeeze out. It was found through extensive experience that no bush material except for austenitic manganese steel (AMS) could survive the service and this material is still used to this day.

Problem 4.3

Why is AMS suitable for the bucket bush?

Answer 4.3

The austentic microstructure of the material is extremely tough and can withstand an enormous amount of mechanical abuse without cracking or breakage. When deformed by stresses above the yield stress, its work hardens rapidly, increases in strength, and only cracks when its considerable ductility is exhausted. Under compression loads, the work-hardened material can attain hardness levels of 700 HV, as Case Study 3 in this chapter has shown.

Problem 4.4

Your Own Work

(a) The hardness of a steel is often used interchangeably with its tensile strength. Why is this practice used and what is the relationship between hardness and tensile strength?

(b) In a hardenable steel, what does 'limiting ruling section' mean?

(c) In Case Study 3 in this chapter, the microstructure of the pin, even near the surface, was mainly bainite with some ferrite. Use the continuous cooling transformation (CCT) diagram for the steel to explain the transformation kinetics it had undergone, assuming an oil quench.

(d) In (c) above, using a water quench would very likely eliminate the soft and undesirable ferrite phase, and further, water quenching would be much cheaper than an oil quench. What are the disadvantages of a water quench?

Appendix 4.1: Tin Dredge Wheel Pinion Failure

Data for Pinion

1. Spur gear, 20° pressure angle
2. No. of teeth (T)   = 25, mating on 152 teeth
3. Speed (N)   = 28.73 rpm
4. Pitch diameter (d)   = 27.852"
5. Face width (b)   = 16"

6. Motor power	= 1000 hp
7. Material	= BS 592 Grade C
8. Transmitted tangential load (F_t)	= 35.16 T = 78,758 lb
9. Diametral pitch (P)	= $\dfrac{25}{27.852}$ = 0.8976 tooth/in diameter
10. Pitch line velocity (v_P)	= πNd = 209.49 ft/min

Appendix 4.2: Stresses According to BS 436

Part 1: Calculating Wear

BS 436 specifies that the allowable tangential load per inch face width for wear of suitably lubricated gears should not exceed the figure obtained from the following,

$$\frac{X_c Z S_c}{K}$$

$$\text{i.e.,} \quad \frac{F_t}{b} \le \frac{X_c Z S_c}{K} \, \sigma c$$

or

$$S_c \ge \frac{K F_t}{b X_c Z} \tag{A4.1}$$

where,

b = face width = 16′
X_c = speed factor for wear = 0.41 for 24-hour operation at 28.73 rpm
Z = zone factor = 2.75 for 25 teeth mating on 152 teeth
K = $P^{0.8}$ = $0.8976^{0.8}$ = 0.9172
S_c = surface stress factor of material for BS 592 Gr. C, S_c = 1400

Substituting these values into the right-hand side of Equation (A4.1) gives,

$$\frac{K F_t}{b X_c Z} = \frac{0.9172 \times 78758}{16 \times 0.41 \times 2.75} = 4004$$

Since S_c = 1400 < 4004, the material used (BS 592 Gr. C) is unsafe for wear. A suitable material would be a surface-hardened EN9 steel with S_c = 40,000.

Part 2: Calculating Strength

BS 436 specifies that the allowable tangential load per inch face width for strength should not exceed the figure obtained from the following,

$$\frac{X_b Y S_b}{P}$$

that is,

$$\frac{F_t}{b} \leq \frac{X_b Y S_b}{P}$$

or

$$S_b \geq \frac{P F_t}{b X_b Y} \tag{A4.2}$$

where,

S_b = bending stress factor for material. For BS 592 Gr. C, S_b = 19,000
X_b = speed factor for strength = 0.45 for 24-hour operation at 28.73 rpm
Y = strength factor = 0.76 for 25 teeth mating on 152 teeth

Substituting these values into the right-hand side of Equation (A4.2) gives,

$$\frac{P F_t}{b X_b Y} = \frac{0.8975 \times 78758}{16 \times 0.45 \times 0.76} = 12,919$$

Since S_b = 19,000 > 12,919, the material (BS 595 Gr. C) is safe in strength.

Appendix 4.3: Stresses According to AGMA

Part 1: Calculating Wear

For wear, the AGMA wear equation is given by,

$$\sigma_c \geq C_p \sqrt{\frac{F_t C_o C_s C_m C_f}{C_v d b I}} \tag{A4.3}$$

where,

$\sigma_c\, a_c$ = calculated contact stress number
C_p = coefficient depending on elastic properties of materials

F_t = transmitted tangential load, lb
C_o = overload factor
C_v = dynamic factor
d = pinion pitch diameter, in
b = face width of tooth, in
C_s = size factor
C_m = geometry factor
C_f = surface condition factor
I = geometry factor

Evaluation of quantities

1. For spur gears made of steel, $C_p = 2300$.
2. For a uniform power source driving a load with moderate/heavy shock, $C_o = 1.50$.
3. For commercial spur gears, $C_v = \dfrac{50}{50 + \sqrt{V_p}}$

that is,

$$C_v = \frac{50}{50 + \sqrt{209.5}} = 0.78$$

4. The size factor C_s varies between 1 and 1.25; for the size of pinion used, a factor of 1.20 is taken.
5. For the third reduction gearing in the system, for a pinion with a 16″ face width, $G_m = 1.43$.
6. For a gear ratio of 6.08 and a 25 teeth pinion, $I = 0.122$.
7. For the finish obtained in the pinion, C_f is estimated to be 1.20.
8. Evaluation of σ_c,

$$\sigma_c = 2300\sqrt{\frac{78758 \times 1.50 \times 1.20 \times 1.43 \times 1.20}{0.78 \times 27.852 \times 16 \times 0.122}} = 18,165$$

9. The AGMA specifies that,

$$\sigma_c \geq S_{ac}\sqrt{\frac{C_L C_H}{C_T C_R}} \tag{A4.4}$$

where,

S_{ac} = allowable contact stress number
C_L = life factor

C_H = hardness ratio factor
C_T = temperature factor
C_R = factor of safety

10. As the material hardness of BS 592 Grade C is about 163 HB, $S_{ac} \approx 8000$.
11. For a required life of greater than 10^7 cycles, $C_L = 1$.
12. Assuming that both the pinion and gear have similar hardness, $C_H = 1$.
13. Assuming that the pinion temperature does not exceed 250°F, $C_T = 1$.
14. The factor of safety $C_R = 1.25$ is used to ensure high reliability.

Then,

$$S_{ac}\left(\frac{C_L C_H}{C_T C_R}\right) = \frac{78,000 \times 1 \times 1}{1 \times 1.25} = 62,400$$

Since $\sigma_c = 181{,}651 > 62{,}400$, the pinion is not safe in wear.
To be safe in wear,

$$S_{ac} \geq 181{,}651 \times 1.25 = 227{,}063$$

This can only be achieved with a high-quality carburized case with a hardness of about 700 Hv (60 Rc).

Part 2: For Strength

The AGMA beam strength equation is a modification of the Lewis equations; it applies various empirical correction factors to account for conditions not considered in the Lewis equation. The equation is written,

$$\sigma_t = \frac{F_t K_o P K_s K_m}{K_v b J} \tag{A4.5}$$

where,

σ_t = calculated stress at root, psi
F_t = transmitted load, lb
K_c = overload correction factor
P = diametral pitch
K_s = size correction factor
K_m = load distribution correction
K_v = dynamic factor
b = face width, in
J = geometry factor

Evaluation of quantities

1. For a uniform power source driving a load with a moderate/heavy shock, $K_o = 1.25$.
2. For spur gears, $K_s = 1$.
3. For less rigid mountings and less accurate gears with contact across the full face, and for $b = 16''$, $K_m = 2.0$.
4. For spur gears finished by shaping,

$$K_v = \frac{50}{50 + \sqrt{V_p}} = \frac{50}{50 + \sqrt{209.5}} = 0.78$$

5. The geometry factor depends, amongst other factors, upon the accuracy of the gear teeth.
6. For a 20 spur gear with 25 teeth, J varies between 0.26 and 0.35. A mean value of 0.30 is chosen.
7. Evaluation of σ_t,

   Substituting all the quantities in Equation (A4.5) gives,

$$\sigma_t = \frac{78758 \times 1.25 \times 0.8976 \times 1 \times 2}{0.78 \times 16 \times 0.30} = 47,204 \text{ psi}$$

8. Maximum allowable design stress:
   The value of σ_t calculated should not be greater than the maximum allowable design stress, given by AGMA as,

$$\sigma_t \leq S_{ad} = \frac{S_{at} K_L}{K_T K_R} \tag{A4.6}$$

where,

S_{ad} = maximum allowable design stress, psi
S_{at} = allowable fatigue bending stress for material, psi
K_L = life factor
K_T = temperature factor
K_R = factor of safety

Now, for BS 592 Grade C, S_{at} can be taken to be 19,000 psi. K_L can be taken to be 0.8 for a life exceeding 10^8 cycles. K_T can be taken as 1 for operating temperatures less than 250°F. K_R is taken as 1.50 for high reliability.

Then, for the pinion made of BS 592 Grade C,

$$S_{ad} = \frac{19000 \times 1}{1 \times 1.50} = 12{,}667 \text{ psi}$$

Since $\sigma_t = 47{,}204 > S_{ad}$, the pinion is unsafe in fatigue.

A suitable material must have a S_{at} value not less than $47{,}204 \times 1.5/0.8 = 70{,}806$ psi. This can only be achieved by having a carburized case with hardness of 700 Hv.

References

General References

ASM Handbook, Vol. 9, *Metallography and Microstructures*, Ed., G.F. Vander Voort, ASM International, Ohio, 2004.

ASM Handbook, Vol. 11, *Failure Analysis and Prevention*, Ed., W.T. Becker and R.J. Shipley, ASM International, Ohio, 2002.

ASM Handbook, Vol. 12, *Fractography*, R.L. Stedfeld et al., ASM International, Ohio, 1987.

ASM Handbook, Vol. 13A, *Corrosion: Fundamentals, Testing, and Protection*, Ed., S.D. Cramer and B.S. Covino, Jr., ASM International, Ohio, 2003.

ASM Handbook, Vol. 19, *Fatigue and Fracture*, S.R. Lampman et al., ASM International, Ohio, 1996, pp. 596–597.

Specific References

1. *Specification for Casing and Tubing*, API Specification 5CT, 8th ed., 2005.
2. Dugstad, A., Mechanism of Protective Film Formation during CO_2 Corrosion of Carbon Steel, Corrosion, Paper #031, NACE International, Houston, TX, 1998.
3. Kermani, M. and A. Morshed, Carbon Dioxide Corrosion in Oil and Gas Production—A Compendium, *Corrosion* Vol. 59, No. 8, 2003, pp. 659–683.
4. Fosbøl, P.L., E.H. Stenby, and K. Thomsen, *Carbon Dioxide Corrosion: Modelling and Experimental Work Applied to Natural Gas Pipelines*, Technical University of Denmark, Engineering Center for Energy Resources, 2007.
5. Stokes, A., *Gear Handbook: Design and Calculations*, Society of Automotive Engineers, 1992.
6. Chironis, N.P., *Gear Design and Application*, New York: McGraw-Hill, 1967.
7. Blau, P., ASM Handbook, Vol. 18, *Friction, Lubrication, and Wear Technology*, ASM International Handbook Committee, Ohio, 1992.

5

Electrical Equipment Failures

5.1 General Introduction

5.1.1 Overview of the Electrical Network

The electrical network comprises the power plants, the transmission system, the distribution system, and finally the end users.

Power plants: These plants generate electricity, traditionally by means of turbines or reciprocating engines that convert mechanical energy to electrical energy in an electrical generator in accordance with Faraday's law of induction. The turbines may be powered by wind, water, steam, or gases that are produced by the direct combustion of hydrocarbon fuels. Reciprocating engines are rarely used for large-scale generation due to high fuel costs, but they are useful as small flexible units in remote locations, for example in the states of Sabah and Sarawak in East Malaysia. At the present time, photovoltaic cells are available, which use solar radiation to produce electricity directly, and which can be harnessed and connected to the electrical grid. Wind, water, and solar energy are so-called green energy sources whose usage will increase in the near future because of fears of carbon dioxide pollution from coal plants and its effect on global warming.

The transmission system: The electricity produced at power plants needs to be transmitted over long distances and wide areas using conducting cables. The cables may be overhead, underground, or submarine, depending on the terrain traversed. The conductor materials used are copper (Cu) and aluminum (Al), whose electrical resistance (R), though low, are nevertheless finite and will cause heat losses (H). These losses, as specified in Ohm's law ($H = I^2R$), are a function of the square of the current, I. For a constant amount of power transmitted, since power (P) is a direct function of (I) and the voltage (V), it makes sense to reduce I to low values by increasing V. Hence, power stations would have substations with step-up transformers to increase the output voltage (say, 32 kV) to the transmission voltage, which may be as high as 1000 kV. The reduction of H is not the only benefit, as lower currents will permit smaller, lighter, and cheaper conductors to be used. In addition, lighter conductors will reduce handling and installation costs, and in overhead lines they will require both lighter and cheaper supporting structures for the same power transmitted (transmission towers and ancillaries).

The distribution system: The transmission voltages are stepped down in stages in various transmission substations and to the distribution substation at the final end to between 4 and 35 kV. The primary feeders from the distribution substation are then routed to light/medium industries or to secondaries for stepping down the voltage further for commercial and residential users.

5.1.2 Insulation Integrity and Breakdown

The insulators used for electrical equipment may be gaseous, liquid, or solid. Examples of gaseous insulators are air, electronegative gases such as sulfur hexafluoride (SF_6) and vacuum ($<10^{-3}$ mbar). Examples of liquid insulators are natural, mineral, and synthetic oils. Examples of solid insulation are paper, wood, rubber, ceramic, and polymeric compounds. A modern polymeric insulation for conducting cables is XLPE (cross-linked polyethylene), which can withstand sustained usage at 90°C. An electrical insulator is a poor conductor of electricity; a dielectric is an insulator that is capable of being polarized for functions other than preventing conduction. For the purposes of this text, the terms *insulator* and *dielectric* will be used interchangeably.

An electrical network can simply be viewed as a system of interlinking and insulated conductors that have various potentials, all higher than the ground (or earth) potential, which is at zero V. Generally, if two conductors of different electric potentials are surrounded by electrical insulation and placed in a uniform electric field, there is a tendency for the electric charge from the conductor with the higher potential to flow to the conductor with the lower potential to equalize the differences in the potentials. If the breakdown strength of the insulator is not exceeded, there will be minimal electric charge flow and very low resultant current. However, if the potential differences exceed the breakdown strength of the insulator, disruptive discharge (DD) would occur between the two conductors (or electrodes) often resulting in an explosion and fire. Proper insulation design is required in electrical equipment for the prevention of DD, which has been given different names for different situations. When it occurs across a gaseous or liquid medium, it is called a *sparkover*; over the surface of solid insulation in gaseous/liquid media, it is called a *flashover* (Case Study 1); and through solid insulation, it is called a *puncture*. If the electric field is highly non-uniform, partial breakdown can occur at locations of high stress but not immediately cause DD.

Professional engineers design electrical equipment with service electric stresses much lower than the breakdown strengths of the insulation [1], but nevertheless, various phenomena can occur that magnify service stresses and reduce breakdown strengths, which can lead to a full breakdown or partial breakdown of the insulation. Partial breakdown and the attendant degradation of the insulator may lead to a full breakdown if such conditions are not detected and rectified. Service stresses can be magnified by switching overvoltages, electrical faults, and lightning strikes. Dielectric degradation of gaseous/liquid media can occur through contamination with water and other species, and from age and thermal degradation and partial breakdown processes. In solid dielectrics, volume or surface degradation can occur; volume degradation can be caused by age, thermal effects, and electrical or water 'treeing'; surface degradation can be caused by the ozone that is formed during partial breakdown, which is irreversible unlike contamination by water, dust, and foreign ionic species.

5.1.3 Switching and Protection

At every stage in the delivery of power from substations, switchgear (circuit breakers, CB) will be installed to make or break currents. Each CB will feed power to a particular downstream location. The ELCB (earth leakage circuit breaker) installed in the incoming circuit in residential households is an example of a simple air-break switch. Air is only a weak insulating medium but is quite sufficient to extinguish the low-energy arcs formed at residential voltages (230 V and less). On the other hand, at transmission and distribution voltages, better insulators that are also more efficient in extinguishing the high-energy arcs formed are needed to allow equipment to be made more compact. Various gaseous or liquid media are available, depending upon the potentials involved; these could be an electronegative gas such as SF_6 (sulfur hexafluoride), vacuum at pressures <10^{-3} mbar [2], or oil. Various oils can be used, natural, mineral, or synthetic but all have a risk of oil fires in the event of a sustained fault.

It is impossible to completely prevent the occurrence of electrical, mechanical, or structural faults in the electrical network, however good the engineering were to be. Nonetheless, the consequential damage from electrical faults can be minimized by a good protection system. The function of the system is to cut off supply to a localized area if there were a fault at that location. It does so by activating the controlling CB when excessive currents occur over a pre-determined length of time. The system should not be so sensitive as to trip too easily (nuisance tripping) or so insensitive as to allow extensive damage to occur before tripping occurs. Generally, the substation would have an outgoing relay to protect each feeder and also an incoming relay to protect the transformer, in case the feeder relay malfunctions. There may also be a global system, which monitors the health of the whole network, which may be set to act automatically, or be activated manually to prevent large-scale power failure. The global system would be able to monitor the presence of voltage dips, swells or spikes, electrical faults and distortion of waveform, collectively called *power quality*. Other parameters can also be measured, depending upon the sensors and instrumentation installed, for example, the real-time temperatures of underground cables (higher temperatures than normal would indicate overloading). In addition, lightning detectors can be installed over any given location, to provide a real-time map of lightning strikes at the monitored locations, which will provide information on whether any fault has been precipitated by lightning. It should be understood that a lightning strike on an overhead line can cause traveling waves (spikes, transients) in the line that can affect equipment tens of kilometers away.

5.1.4 Failure Analysis of Electrical Equipment

The same principles and procedures used for non-electrical components also apply for electrical components, although the expertise required and the types of testing differ. The investigators would need to have a good knowledge of electrical circuit theory, electrical equipment and instrumentation, and electrical failure modes, given the many things that can go wrong with engineered components. This subject matter is wide and it will require a team of FIs schooled in different electrical disciplines to perform the tasks efficiently.

As mentioned in Chapter 1, the main electrical failure mechanisms are disruptive DD and ohmic heating. Ohmic heating damage is not detected by the protection relays in a CB and often proceeds to a stage when DD is triggered, due to combustion damage of the insulation and the subsequent contamination of the surrounding atmosphere. The identification of the proximate causes then becomes difficult and clues need to be obtained from power quality records or lightning records. Bolted or spring-loaded joints are often sites for ohmic heat damage, due to poor design, installation, maintenance, or poor working conditions.

An example is given here to illustrate the results of an electrical failure investigation into the explosion and fire of an oil-filled transformer. Figure 5.1 shows the fire that ensued after the initial explosion, and Figure 5.2 shows arcing marks on an internal steel structure, which was discovered after the fire had been put out and the remnants of the oil removed. All other damage on the outside of the tank was established to have been caused by fire. There were no records to show that periodic oil testing and oil filtering were performed; these are required to ensure that the dielectric properties of the oil remained within specifications. Power quality records did not show the presence of abnormalities prior to the incident and lightning records did not show any untoward activity. It was thus concluded that the proximate cause was sparkover across the degraded oil from a conductor under normal load and that the root cause was a lack of maintenance of the insulating oil.

FIGURE 5.1
(See Colour Insert.) Fire at an oil-filled transmission transformer, caused by an internal sparkover.

FIGURE 5.2
Circles enclose the arcing marks on the internal steel structure of the transformer shown in Figure 5.1, caused by a sparkover from a conductor, across the contaminated insulating oil.

Three case studies are presented in this chapter:

Case Study 1: Failure of an On-Load Tap-Changer (OLTC)
Case Study 2: Failure of Two Induction Motors
Case Study 3: Damage to a Transformer Due to Water Ingress

5.2 Case Study 1: Failure of an On-Load Tap-Changer

5.2.1 Tap-Changing Principles

The on-load tap-changer (OLTC) is a device fitted to a transformer to regulate the magnitude of its output voltage within small limits. This is necessary because the output voltage of power transformers are affected by the load drawn and the supply voltage must be kept to within set limits to allow proper functioning of electrical equipment connected to the supply. Regulation is effected by varying the output turns ratio slightly, either by an off-load or an on-load process; the latter process allows changes to be made without power disruption and is much preferred. The turns ratio is changed in discrete steps, which means that the current must be switched from one physical contact to another, a process known as *tap changing*. In a tap change, a new connection is made before the old one is released, by the use of switching devices, of which several different designs exist. The equipment in this case was an ABB Type UZ[3], where the selector switch consists of fixed contacts and a moving contact system. The moving contact comprises a main contact with switching contacts (diverter switches) on either side, each of which contains a high-resistance, transition resistor wire (TRW). The fixed contacts are mounted in an epoxy moulding and only the contact arms are visible; these arms are broad enough to accommodate one main contact and one switching contact at a time. In a fully switched condition, the main moving contact will bear on the assigned fixed contact, with the switching contacts on either side clear of the main contact.

To explain the switching sequence in the ABB OLTC, we first designate the two fixed contacts as FC1 and FC2, the main moving contact as MMC and the switching contacts on the right and left sides of the MMC as SCR and SCL, respectively. MMC, SCR, and SCL are all mounted as one unit on a shaft and rotate in unison. Initially, only MMC bears on FC1, in a central position; a switching sequence from FC1 to FC2 will require MMC to be rotated in a clockwise direction. The sequence of movements will then be:

1. SCL will move and make contact with FC1; MMC and SCL will share the load current.
2. MMC will break contact with FC1 and only SCL with the transition resistor will take the load current. It must only remain momentarily in this position or overheating of the resistor and failure will ensue.

3. SCR will make contact with FC2 while SCL still remains on FC1. The load current is shared between SCL and SCR and circulating currents are limited by the transition resistors.

4. SCL breaks from FC1 and SCR will now carry the load current. Again, it must only remain momentarily in this position or overheating of the resistor and failure will ensue.

5. MMC makes contact with FC2; MMC and SCR will share the load current.

6. Finally, SCR breaks from FC2 and only MMC carries the load current. The tap change is complete.

5.2.2 Background Information

The OLTC was fitted to a 132/11 kV transformer, which tripped during operation. Tripping occurred at the transformer differential relay at both the yellow and blue phases and at the OLTC pressure relay; tripping was caused by an internal DD. Operation and maintenance records were not available so it is not possible to say if they had conformed to the manufacturer's instructions. No further information was available to allow the condition of the OLTC before failure to be evaluated, so investigation had to be based mainly on physical examination.

5.2.3 Visual Examination

The OLTC had 12 contact blades numbered from 1 to 12/12b; blade 12 is continuous with 12b. Figure 5.3 shows the internal components after the cover was removed; the yellow phase (Yph) with broken resistor wires was at the centre, the red phase (Rph) was on the left, and the blue phase (Bph) on the right. The ellipse encloses part of the broken resistor wire from the SCR; the moving contacts are shown resting on contact blade 1. Note that this was not the actual position after failure, as the moving contact positions had been changed during initial examination before opening up. Figure 5.4 shows a view of the Yph after partial cleaning, with the MMC and SCR sitting on the fixed contact blade 12. This was said to be the actual resting position of the moving contacts after failure, before their position was changed. In this position the broken region of the TRW was directly opposite to the contact feet 12b and 1. Arcing damage was evident on the terminal to which the resistor wire was brazed.

Figure 5.5 shows the broken resistor wire of the SCR at the Yph, with dark, carbon deposits on the whitish insulating bobbin, which had been created by a DD to the wires. Parts of the wire that had been damaged and darkened by arcing are indicated by arrow A and the damaged wire terminal from Figure 5.4 is indicated by arrow B. Figure 5.6 shows that the brass bracket and steel bolts attached to contact 1 contained arcing marks. A tracking mark

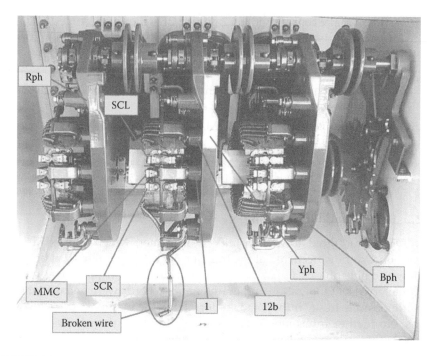

FIGURE 5.3
(See Colour Insert.) View of the internal components of the OLTC after removing the cover of the tank. The yellow phase (Yph) is at the centre, the red phase (Rph) is to its left side, and the blue phase (Bph) is to its right. The moving contacts MMC and SCR are seen to be resting on contact 1. The ellipse encloses part of the broken resistor wire from the SCR.

was present on the surface of the epoxy moulding, which joined to contact 12. The surface of the moulding at this region contained a layer of dark deposits, likely sludge from slow degradation of the oil. Figure 5.7 shows a full view of the tracking mark in Figure 5.6; it was evident that the track was essentially continuous between contacts 12b and 1. Arcing damage was present on the brass brackets for securing the fixed contact blades. The dark deposits on the surface of the moulding had been cleaned up before this photograph was taken. Figure 5.8A shows the damaged SCR resistor wire after cleaning up. Note that only the part adjacent to the wire terminal, over about 25% of its total length, had been damaged. Figure 5.8B shows a magnified view of pieces of wire, broken off by arcing damage and darkened. Figure 5.8C shows a magnified view of the wire terminal, containing arcing damage. Figure 5.9 shows the Rph moulding, at the area between contacts 10/11 and 12/12b, also containing darkish sludge deposits on the surface near to the feet of the contacts.

5.2.4 Laboratory Examination

Dissolved gas analysis (DGA): This was performed by the client after the failure and an analysis of the gases produced using the

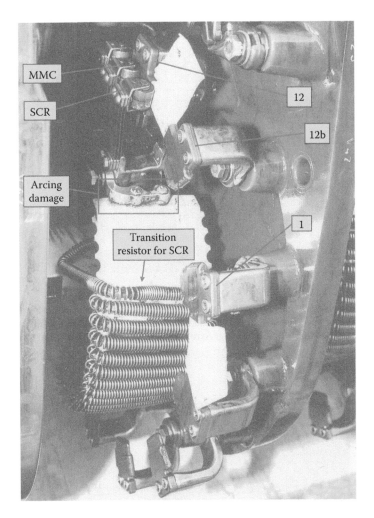

FIGURE 5.4
(See Colour Insert.) View of the Yph after partial cleaning, showing the MMC and SCR sitting on the fixed contact blade 12. This was said to be the resting position of the moving contacts after failure. In this position it can be seen that the broken region of the transition resistor wire was directly opposite to the contact feet 12b and 1. Arcing damage was also evident on the terminal to which the resistor wire was brazed (boxed).

Duval triangle indicated that there was a low-energy electrical discharge. The dielectric strength of the insulating oil was found to be below specifications.

SEM/EDS of the broken SCR resistor wire: EDS detected the main elements in the wire to be copper, manganese, nickel, and aluminum. The wire was a proprietary grade of manganin, an alloy which is commonly used for resistance wires. Under the SEM, the surface of the wires was seen to be covered with solidified melt, with a typical

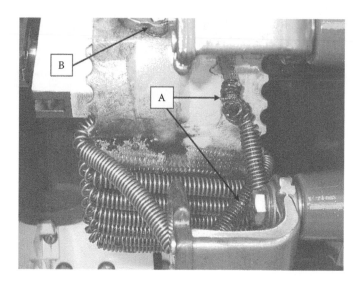

FIGURE 5.5
View of the broken resistor wire of the SCR at the Yph; the dark deposits on the whitish insulating bobbin had been caused by DD to the wires. Arrow A points to parts of the wire that were damaged and darkened by arcing. Arrow B points to the damaged wire terminal from Figure 5.4.

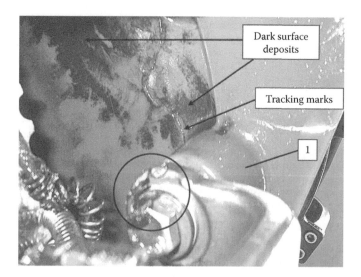

FIGURE 5.6
The brass bracket and steel bolts attached to contact 1 of the Yph contained arcing marks (circled). A tracking mark was present on the surface of the epoxy moulding, which joins to contact 12. The surface of the moulding at this region contained a layer of dark deposits.

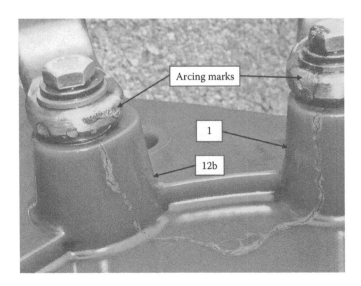

FIGURE 5.7
View of the tracking marks between contacts 12b and 1, which were partially illustrated in Figure 5.6. Arcing damage is clearly visible on the brass brackets that are used for securing the fixed contact blades. The dark deposits on the surface of the moulding had been cleaned up before this photograph was taken.

appearance of damage caused by a fast pulse of heat (Figure 5.10). A fresh fracture surface made at room temperature, under high magnification, revealed a ductile failure with dimples, which were formed by microvoid coalescence (Figure 5.11). On the other hand, the original fracture surfaces contained intergranular facets, typical of fractures at high temperatures (Figure 5.12).

Other tests: An X-ray was performed on the Yph moulding to ascertain if there could have been any damage at the encased conductors that could have contributed to the failure but no signs were seen.

5.2.5 Summary of Results and Analysis

1. The DGA predicted the occurrence of low-level DD, which confirms the physical evidence. However, this is of little use to predict the prior existing conditions that had led to the DD.

2. The TRW of SCR had suffered extensive damage as a result of sparkover from another conductor. One terminal to which the manganin wire was connected also contained arcing damage. The fractures at the manganin wire had occurred at high temperatures. These features could have been caused in two ways; firstly, from a pure sparkover from another conductor, and secondly, from initial

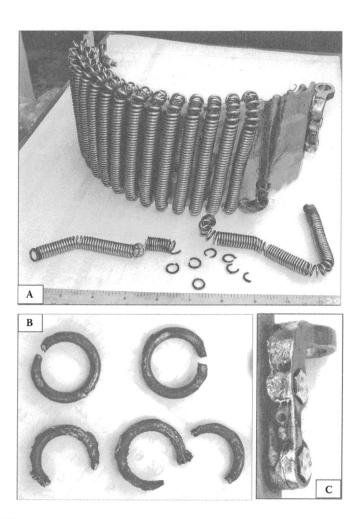

FIGURE 5.8
(A) Shows the damaged SCR resistor wire after it had been cleaned up. Note that only the part adjacent to the wire terminal (T) had been damaged, about over 25% of its total length. (B) Shows a magnified view of pieces of wire, broken off and darkened by arcing damage. (C) Shows a magnified view of the wire terminal, containing arcing damage.

overheating, that led to wire breakage and then arcing between the open-circuited wires.

3. The TRW of the SCR was located opposite to contacts 12b and 1 at the time that tripping occurred. In this position, a sparkover could have occurred between the TRW and contacts 12b or 1.

4. There were arcing marks on the brackets holding the metal contacts at contact feet 12b and 1. The arcing marks on the brackets had likely been caused by a flashover between 12b and 1, as well as between

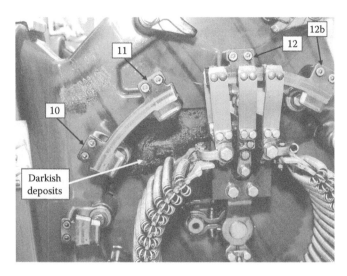

FIGURE 5.9
Rph moulding, at the area between contacts 10/11 and 12/12b, also containing darkish deposits on the surface near to the feet of the contacts.

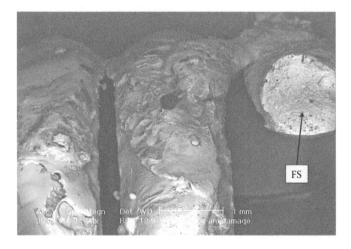

FIGURE 5.10
View of an SEM photograph of some broken wires. The surface of the wires was covered with solidified melt, typically caused by a fast pulse of heat. The fracture FS was freshly cleaned up for investigation purposes. Original magnification: 24×.

12b or 1 and TRW. Continuous tracking marks were also present on the epoxy surface of the moulding between these contacts. Tracking had been facilitated by the accumulation of sludge deposits on the epoxy surface between contacts 12b and 1, which were partially conductive in nature.

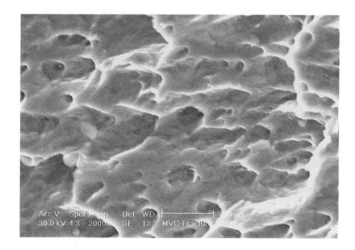

FIGURE 5.11
A scanning electron microscope (SEM) photograph of the fracture surface FS from Figure 5.10, showing a fully dimpled appearance typically caused by MVC. Original magnification: 2000×.

FIGURE 5.12
An SEM photograph of an original fracture surface, showing predominantly intergranular characteristics. Original magnification: 851×.

5.2.6 Most Probable Sequence of Failure and Conclusions

1. Of all the possible damage mechanisms enumerated in Section 5.2.5 above, all were very fast occurring events except for the tracking phenomenon in item 4. The tracking is then deduced to have been the precursor to the failure, as it could not have been subsequently caused if any of the other mechanisms had first occurred. The most likely sequence of failure would then be:

- Long-term coking had led to large amounts of partially conductive sludge deposits on the epoxy surface between contact feet 12b and 1. This had promoted the formation of an (electrical) track between the two contact feet, which culminated in a flashover between the two contacts.
- The flashover generated high temperatures and large amounts of conductive gases, which allowed a sparkover to occur between contacts 12b or 1 and the TRW of the SCR. Later, a DD involved other locations not described here.
- Damage to the TRW was a consequential event.

2. If periodic DGA had been performed during maintenance and the results properly analyzed, the poor condition of the OLTC could have been detected and rectified to prevent failure.

5.3 Case Study 2: Failure of Two Induction Motors

5.3.1 Background Information

This case involved two out of three similar induction motors that were installed at about the same time, at the same location, and subjected to the same type of service (M1, M2, and M3). They were used to drive compressors that handled chilled air and glycol in a district cooling system in a large shopping mall. The equipment had been under continuous service for about four years and in the process were started and stopped twice a day. Starting was effected via a programmable logic controller (PLC) to control the inrush current to give a 'soft' start. The second and third motors (M2 and M3) had failed within a few weeks of each other. Initial reports by the client indicated that both motors had suffered from damage at the bearings, the stator, and the rotor. Apparently, prior to the failure, abnormal noises and vibrations were detected, the causes of which could not be identified. There were also several occasions when the current intake was excessive. These were indications that some operating conditions were unsatisfactory, but the motors needed to be kept running to enable the temperatures in the mall to be kept to comfortable levels, as otherwise business would be badly affected.

Basic specifications and details of the motors were as follows:

1. Make: Internationally reputable brand
2. Type: 2-pole, 3-phase, 3.3 kV, 3000 rpm, 1250 hp (932 kW), and 6350 lb (2880 kg)
3. Insulation: Type F (maximum 155°C at ambient temperature of 40°C)
4. Enclosure type: WPI (weather protected type I with no air inlet filter fitted)

5. Full load current: 186 A, service factor 1.15

6. Rotor cage: One-piece, cast in, Al alloy material; made up of 39 bars with end rings on either end

7. Automatic tripping: Set at motor current of 186 × 1.15 = 212.9 A

Note that no sensors were installed to monitor temperatures at the stator or the bearings, or shaft vibration levels. If stator temperature sensors had been installed, they would give an alarm at 155°C and activate tripping at 170°C. There were no complaints of excessive voltage variations in incoming power, and an inspection of voltage quality records for the month preceding the failure showed that variations were within 3% of the specified 11 kV, which was considered to be normal. The rotor cage material was very likely an Al-Si (aluminum-silicon) casting alloy with a eutectic temperature of about 577°C, which is the temperature at which it will start to melt when heated up.

5.3.2 Basic Construction and Characteristics of Induction Motors

The induction motor consists of a stationary stator and a rotating rotor (or armature); both components contain conductor coils embedded in slots in their respective laminated iron cores. The stator is made in pole pairs and the motor speed is dependent upon the number of poles. For example, a 2-pole motor with a 50 Hz power supply has a synchronous speed of 3000 rpm, and a 4-pole will have a speed of 1500 rpm. When an alternating current flows through the stator coils, a rotating electromagnetic field is generated, which induces secondary currents in the rotor coils. These currents interact with the rotating magnetic field and make the rotor rotate and drive the connected equipment. The voltage applied to the stator coils is high and hence they must be properly insulated to prevent arcing between coils, between the coils and the earthed frame, and between the coils and the rotor. On the other hand, the emf (electromotive force, V) induced in the rotor is low and the rotor coils do not have to be insulated against the iron core. The rotor coil has one end ring (ER) at either end of the iron core and the circumferences of the rings are mechanically and electrically connected by bars, to form the shape of a squirrel cage, hence the commonly applied term, 'squirrel cage rotor'. The bars have to be good conductors and must be robust enough to withstand long service at less than ideal conditions. They are normally made of good conductors such as Al or copper. Al bars may be fabricated or cast into the core from the molten state, whereas Cu bars are normally fabricated.

The currents induced in the rotor cage depend upon the difference in speed (slip) between the stationary stator and the moving rotor; a higher slip will create higher currents. Hence, during starting, when the slip is 100%, the stator and rotor coils will be subjected to very high currents, up to nine times the maximum load current. For this reason, the starting process of induction motors must be closely controlled to prevent overheating and damage (soft start by PLC control). The rated insulation class determines the temperature to which the insulation can be subjected. For example, if Class F insulation is used, it can be heated up to a maximum of 155°C continuously. Generally, continuous operation at the maximum allowable rated temperature will give an average insulation life of 20,000 hrs. Increasing the temperature by every 10°C will reduce life by 50% and conversely, reducing temperature by 10°C will increase life by 100%. Sensors are often installed for condition monitoring purposes, to check for excessive temperatures at the stator or bearings, or excessive vibrations at the bearings. Early indications of the above, if heeded, can prevent serious damage.

5.3.3 Visual Examination

Due to the urgency to repair the affected motors, dismantling works had already commenced before the failure investigation began. The simple items like worn bearing bushings had already been sent for repairs, and examination was confined to only the electrical parts. However, the electrical parts were the ones that first failed and the investigation was able to identify the proximate and root causes convincingly.

Figure 5.13 shows a view of a similar motor in the installed position, coupled to a compressor. The inlet air box was located on the top of the motor, with two outlet air boxes on either side. The inlet box was only fitted with coarse wire mesh as a filter, in accordance with WPI specifications. The installed location was not conducive to easy maintenance and removal for repairs

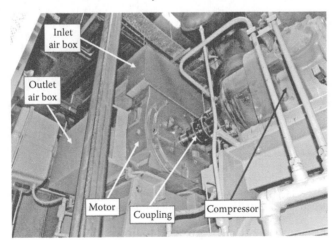

FIGURE 5.13
View of the motor in the installed position, coupled to a compressor. The inlet air box is located on top of the motor, with two outlet air boxes on either side. The inlet box is only fitted with coarse wire mesh as a filter, in accordance with WPI specifications. The motor weighs 6350 lb (2880 kg). The installed location was not conducive to easy maintenance.

FIGURE 5.14
(See Colour Insert.) The M2 stator, as viewed from the drive end.

FIGURE 5.15
(See Colour Insert.) Close-up view of the M2 stator coils as viewed from the drive end; large amounts of melted and re-solidified groundwall insulation were present.

would be a major operation. Figure 5.14 shows the dismantled M2 stator, viewed from the drive end and Figure 5.15 shows a close-up view of the surface of the stator coils containing large amounts of melted and re-solidified groundwall insulation. The surface temperatures had far exceeded the specified over-temperature tripping level of 170°C but the condition was not detected, as there were no stator temperature sensors installed.

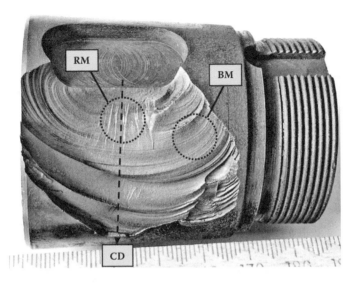

FIGURE 1.7
Fatigue failure initiating from a sharp edge at the milled keyway in a tapered shaft. Arrow *CD* shows the crack direction, area *BM* the beach marks, and area *RM* the ratchet marks on the fracture surface, formed during Stage II crack growth.

FIGURE 1.10
Flashover at an electrical substation, compounded with protection failure, leading to a raging fire and total loss of the electrical switchgear.

FIGURE 1.11
Ohmic heat damage at a loose electrical joint, leading to melting, joint separation, arcing, and total failure.

FIGURE 1.12
Corrosion under insulation (CUI) of galvanized mild steel spacers supporting the outer, stainless steel skin of a very large, rock wool insulated roof, caused by rainwater entry during installation.

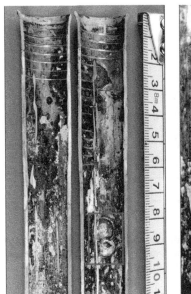

FIGURE 1.13
Microbial corrosion beneath barnacle bases in an aluminum brass, heat exchanger tube with seawater flowing inside; the boxed area at the left figure is magnified at the inset on the right.

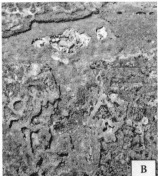

FIGURE 2.1
(A) Site photographs of CO_2 attack on API 5CT, Grade L-80, production tubes from an offshore oil well. (B) Shows a close-up view of a typical attacked location with darkish $FeCO_3$.

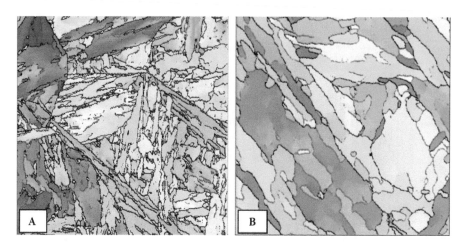

FIGURE 2.3
EBSD-IPF images of ASME A335, P92 steel. (With permission from Guat-Peng Ng et al. [1].)
(A) Shows the normal fine-grained, tempered martensite phase required for creep service at
elevated temperatures. (B) Shows that ferrite with inferior creep strength had formed after
treatment just below the A_{C1} temperature.

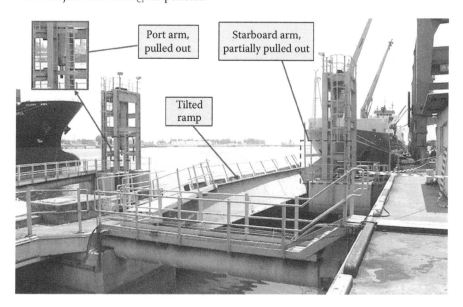

FIGURE 3.12
Photograph showing the tilted ramp with its port side submerged. The port hydraulic piston
arm had pulled out completely from its securing nut; the starboard arm had partially pulled out.

(a)

(b)

FIGURE 3.17
(a) Port arm; upper part of the male thread, at a location about 90° from Figure 3.16, but with similar characteristics. (b) Port arm; lower part of the male thread, at a location about 90° from Figure 3.16, stuck-on threads covered less than 50% of the area. The cross-sectional areas of the stripped-off threads were variable.

(a)

FIGURE 3.19

(a) View of a sample cut out from the recovered nut; the surface of the hole is covered with rust.

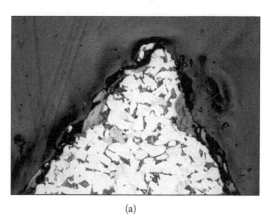

(a)

FIGURE 3.22

(a) At 50× magnification, specimen T1, tooth No. 2 was one of the most complete threads at T1, measured to be 2.2 mm deep from the crown to the root. The arrows point to deep pockets of corrosion on the surface. No signs of deformation were present in the material.

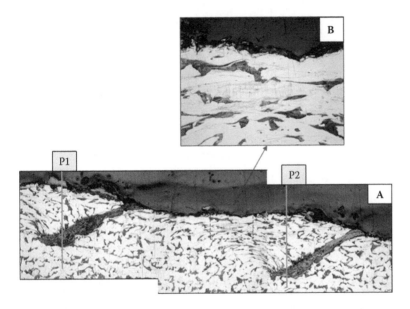

FIGURE 3.26
Specimen B1: (A) Planes P1 and P2 show the locations of adjacent roots of a thread; the thread
has sheared off, towards the right of the figure and caused the underlying material to become
deformed. The sheared surface did not contain deep pockets of corrosion at 25× magnification.
(B) Magnified view of the sheared surface, showing deformed ferrite (light) and pearlite
(darkish) grains at 100× magnification.

FIGURE 3.27
The right girder, showing the broken, central, hinged portion resting on the ground, and the ends still supported on the pier brackets.

FIGURE 3.31
The left girder, stiffened connection box of the male hinge plate; 'BF' stands for bottom flange.

FIGURE 3.38
The right girder, lower connection, with the torn-off male hinge plate attached to the female parts. The female members had been forced up on hitting the ground.

Darkish scale

(a)

(b)

FIGURE 4.1
(a) Close-up view of the damaged tubes, showing highly irregular profiles due to internal pitting attacks. The corroded edges had a darkish-coloured scale (inset), different from the reddish-brown colour of the scale on the unaffected surfaces. (b) Certain tubes appear to have external pits (said to be LS tubes). (c) Close-up view of the external pits from (b): The arrow points to a darkish scale within the pits, but the scale on the unaffected surfaces was reddish brown in colour.

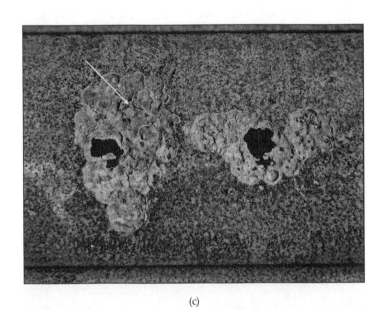

(c)

FIGURE 4.1 (*Continued*)

(a)

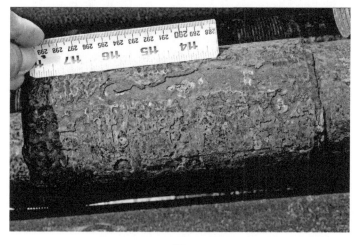

(b)

FIGURE 4.2

(a) The collar with extensive corrosion on its external surface: Attacked areas are sharply demarked from non-attacked areas, and it also has a darkish colouration. (b) Another collar with extensive corrosion on its external surface: Attacked areas are also sharply demarked from non-attacked areas, and it also has a darkish colouration.

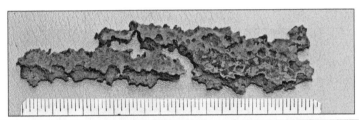

FIGURE 4.8
Loose sample acquired during inspection: Views of the internal surface where the pitting attack had initiated are shown. The surfaces formed by the corrosion attack are highly irregular and contain darkish deposits.

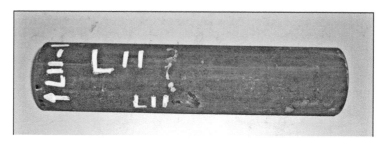

(a)

(b)

FIGURE 4.10

(a) A half-section from sample L11. (b) The internal surface of L11 with well-defined pits containing darkish deposits.

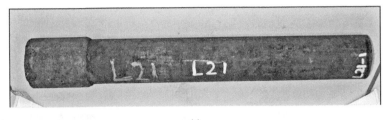

(a)

(b)

FIGURE 4.11
(a) Sample L21 after cutting off specimen L21-2 from the collar. (b) Sample L21: Examination shows that the internal surface of the collar is also badly pitted and covered with darkish deposits. The bright debris on the deposits are metal particles that were formed from the sawing process.

FIGURE 5.1
Fire at an oil-filled transmission transformer, caused by an internal sparkover.

FIGURE 5.3

View of the internal components of the OLTC after removing the cover of the tank. The yellow phase (Yph) is at the centre, the red phase (Rph) is to its left side, and the blue phase (Bph) is to its right. The moving contacts MMC and SCR are seen to be resting on contact 1. The ellipse encloses part of the broken resistor wire from the SCR.

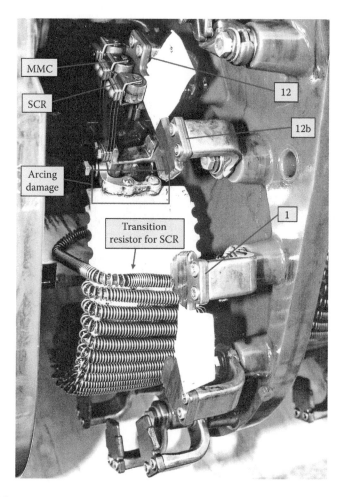

FIGURE 5.4

View of the Yph after partial cleaning, showing the MMC and SCR sitting on the fixed contact blade 12. This was said to be the resting position of the moving contacts after failure. In this position it can be seen that the broken region of the transition resistor wire was directly opposite to the contact feet 12b and 1. Arcing damage was also evident on the terminal to which the resistor wire was brazed (boxed).

FIGURE 5.14
The M2 stator, as viewed from the drive end.

FIGURE 5.15
Close-up view of the M2 stator coils as viewed from the drive end; large amounts of melted and re-solidified groundwall insulation were present.

FIGURE 5.16
View of the dismantled M2 rotor with the drive end (DE) on the right side and the non-drive end (N-DE) on the left. Arrows point to the AI end rings at the DE (ER-DE) and at the N-DE (ER-N-DE).

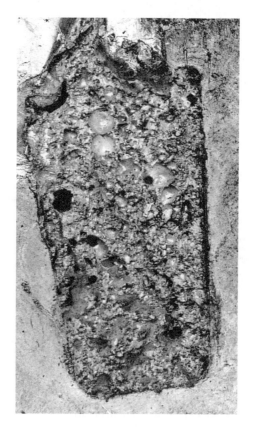

FIGURE 5.21
Close-up view of a typical fracture appearance on the removed ER-DE, containing heavy discolouration and possible arcing damage (at the top right and bottom left), gas voids (the round and oval holes), and micro-shrinkages (all over, fir-tree appearance). The fracture mechanisms were obscured by these features.

FIGURE 5.27
View of the inside of the tank from the Rph turret port; arrows point to darkish colouration on the top surfaces of the insulation boards just below the port.

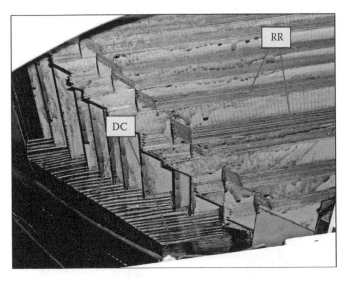

FIGURE 5.28
View of the inside of the tank a little further from Figure 5.27, showing the top edges of the laminations of the steel core to contain reddish rust (RR) and the top surfaces of the end insulator boards to have a dark colour (DC).

FIGURE 5.30
After removing the tank cover, the top of the Rph winding is seen to have large areas of partially dried, darkish-coloured sludge on the horizontal surfaces of the insulation boards.

FIGURE 5.31
The steel core laminations below the Rph turret port contained large but localized areas of reddish rust at the top of the laminations.

FIGURE 5.32
The floor of the tank at the base of the Rph winding contained a fairly large area with old water stains and darkish blobs of sludge (seen from the aft end). A vertical, rectangular support bar was present, which stretched to the top of the winding and helped to secure the parts together.

FIGURE 5.33
The same area as shown in Figure 5.32, but viewed from the port end; the base of the support bar had darkened, swelled, and split along its length. Further, the insulating laminations supporting the winding had warped.

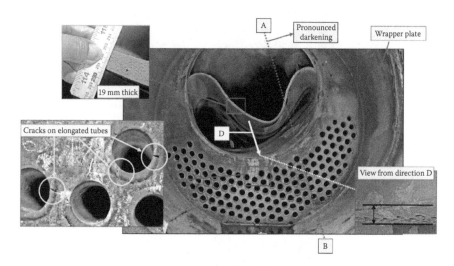

FIGURE 6.3

Boiler conditions at the aft end. (i) The tubeplate and reversal chamber wrapper plate show significant darkening above the horizontal AB, about one-third of the furnace diameter distance from the furnace top. (ii) The ruptured edge of the furnace wall was measured to be about 19 mm thick (inset). (iii) The ends of some second pass tubes had become oval and developed cracks, (iv) when viewed in direction D, the part of the tubeplate close to the furnace tube can be seen to have been pulled inward towards the tube, by deformation of the tube during the incident.

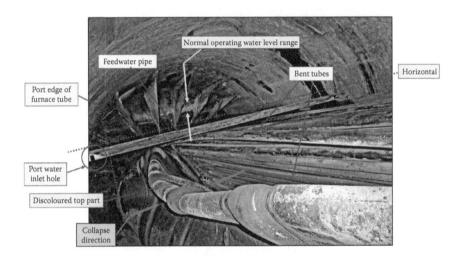

FIGURE 6.4
View of the outside of the furnace tubes, as seen from the aft end: The first pass furnace tube had collapsed by being pinched horizontally on both sides. Some of the second pass tubes had been slightly bent. The top one-third of the furnace tube surface had a much darker colouration than the remainder of the shell. The port water inlet hole was located beyond the port edge of the furnace.

FIGURE 6.5
View of the port, fore side of the (first pass) furnace tube, from the aft end the top one-third of the tube surface is darkened, with a sharp demarcation between the darkened and non-darkened parts.

FIGURE 6.6
Furnace samples F34-6 and F34-12 from between bowling rings 3 and 4; the sample F34-12 from the top of the tube was darkened but F34-6 was not.

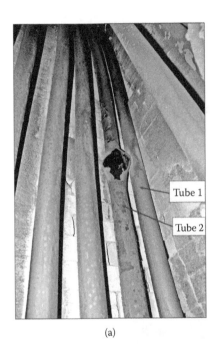
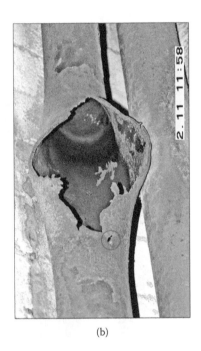

(a)   (b)

FIGURE 6.15

(a) Shows the burst located at the second tube from the starboard side (Tube 2). The parts of the burst tube from about 2 diameters away from the burst did not any show any visual signs of bulging. The surfaces of Tube 2, far away from the burst, and of the adjacent tubes, were covered with a thick, darkish magnetite scale. (b) Shows a close-up view of the burst.

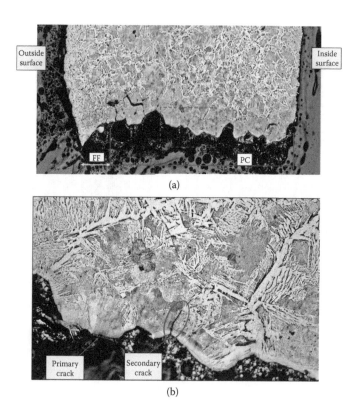

(a)

(b)

FIGURE 6.18

(a) Shows the fracture profile of specimen B1; the fracture profile is made up of the regions 'PC' and 'FF'. Mag. 60×. (b) Magnified view of the centre of the fracture profile (a). The primary crack and secondary crack at this location are seen to be intergranular in nature. The darkish phase is seen to be predominantly very fine pearlite, possibly with some mixed ferrite/carbide phases. The microstructure is coarse. Mag. 300×.

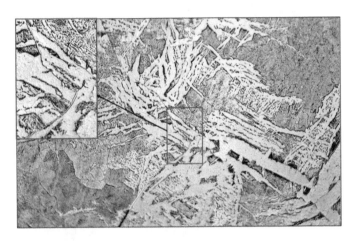

FIGURE 6.19
Magnified view of the secondary crack from Figure 6.18b, showing the presence of a light-coloured, copper phase in the uncracked grain boundaries above the secondary crack. Mag. 750×.

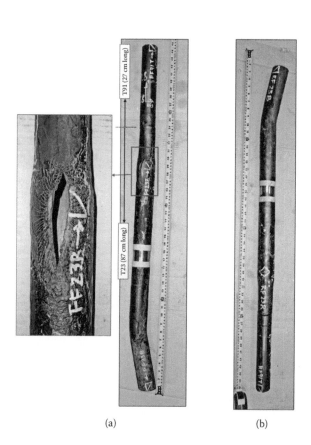

FIGURE 6.22
(a) The front view of the failed location with a close-up view of the rupture. (b) Rear views of the failed location. The failure had occurred at the T23 side. The oxides around the bulge had cracked. The locations of the metallographic specimens FF23B, FF23R, FF91T, RF23B, RF23R, and RF91T are marked on the tubes.

(a)   (b)

FIGURE 7.4
(a) Corroded Z-spacers from rectified bay E1-5 and (b) localized corrosion and pitting attack on the corroded Z-spacer from bay E1-5.

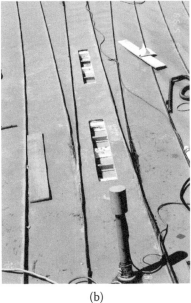

(a)   (b)

FIGURE 7.8
(a) A view of the section made on bay B2-3. The Z-spacers were white-rusted and brown rust was observed to have progressed to a significant level. (b) A rectangular section made on bay D12-6. The spacer was corroded and brown rust was obvious.

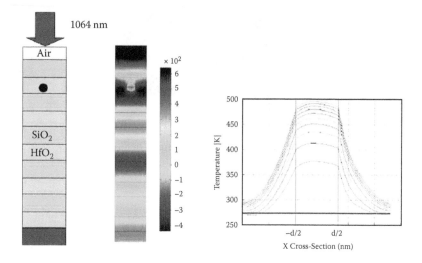

FIGURE 8.3
Light intensification of backscattered and incoming photons over a finite thickness in Commandre et al. in 2010 [10].

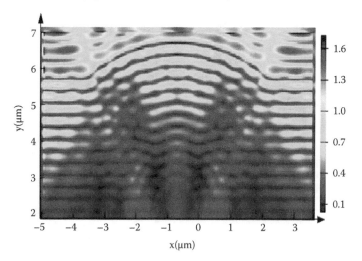

FIGURE 8.4
Light intensification simulation using Lumerica™.

FIGURE 5.16
(See Colour Insert.) View of the dismantled M2 rotor with the drive end (DE) on the right side and the non-drive end (N-DE) on the left. Arrows point to the Al end rings at the DE (ER-DE) and at the N-DE (ER-N-DE).

Figure 5.16 shows the dismantled M2 rotor with the drive end (DE) on the right side and the non-drive end (N-DE) on the left. Arrows point to the Al end rings at the DE (ER-DE) and at the N-DE (ER-N-DE). The drive end (DE) is coupled to the compressor and the other, non-drive end (N-DE) is free. The Al end rings at the drive end (ER-DE) and at the non-drive end (ER-N-DE) were still in place. The surface of the ER-DE had small, bright particles of re-solidified Al on it (Figure 5.17). Since the rotor cage was the only Al component at this location, the particles must have come from it. Temperatures at the cage at certain locations must then have reached the eutectic melting point of 577°C. The surface of the rotor also contained areas that had been polished by rubbing against the stator. This could only have occurred if the bearings had become sufficiently worn out to allow the shaft to become excessively displaced. Figure 5.18 shows that a thick layer of caked-up dirt was present on the internal bore of the ER, an indication of a lack of maintenance.

When the ER-DE was removed, it was observed that the rotor bars had broken at the junction with the end ring (Figure 5.19). Examination of the fractured bars on the rotor revealed the presence of old, severely arc-eroded and oxidized fracture surfaces (Figure 5.20). There was also arcing damage on the last layer of steel core lamination. Examination of a typical fracture appearance on the removed ER-DE using a magnifying glass revealed the presence of heavy discolouration and possible arcing damage, gas voids, and micro-shrinkages (Figure 5.21). The fracture mechanisms were obscured by these features but more detailed analysis was not considered to be necessary. The arcing damage were subsequent events and only occurred after the ER had separated from the rotor bars to create an open circuit. The micro-shrinkages and gas voids were material defects caused by a poorly executed casting process; they would have been detected if non-destructive testing (NDT) had been performed after casting.

The M3 motor was found to contain the same material defects and fracture characteristics as the M2 motor. When the fracture surface of the removed

FIGURE 5.17
View of the ER-DE with small bright particles of re-solidified AI on the surface. The arrow
points to polishing of the rotor surface due to rubbing against the stator.

ER-DE was machined away, a multitude of gas voids were visible all over
the machined face (Figure 5.22). The magnified view in Figure 5.23 reveals
grouped voids exceeding 10 × 5 mm in size. All other comments pertaining
to M2 are equally applicable here.

5.3.4 Summary of Results and Analysis

1. Both the failed motors were of the same make, underwent the same
   type of operation, contained similar defects, and failed in the same
   manner. Both were not fitted with sufficient condition monitoring
   instrumentation (likely to save costs) and were not expertly maintained (very likely due to a difficult working environment). The
   engineering input in these respects left much to be desired.

2. In both cases, fracture had occurred at the rotor cage, between the end
   rings and the rotor bars, which led to overheating of the remaining

FIGURE 5.18
The N-DE of the M2 rotor, showing a thick layer of caked-up dirt on the internal bore of the Al end ring (arrowed).

FIGURE 5.19
The M2 rotor with the ER-DE removed to show that the rotor bars had broken at the junction with the end ring.

FIGURE 5.20
Close-up view of the fractured bars on the rotor, containing old, severely arc-eroded, and oxidized fracture surfaces. Arrows point to the arcing damage on the last layer of the steel core lamination.

bars and arcing between the open-circuited bars. Localized melting of the arced rotor material occurred and caused re-solidified particles to deposit on the external surface of the ER. The internal surfaces of the stator had become overheated, mainly due to proximity to the overheated rotor, causing the groundwall insulation to experience localized melting. The higher than usual stator currents detected before failure was likely a consequence of the rotor bar failures; they would have caused an overall temperature rise at the stator, but no sensors had been installed to monitor such a condition.

3. Failure of the rotor bars would also cause the forces acting on the rotor to become unbalanced and to cause the bearings to become overloaded and to wear out. The abnormal noises reported before the failure occurred were caused by excessively worn bearings, but since vibration sensors were not installed the condition was not positively detected.

4. Large quantities of material defects in the form of micro-shrinkages and gas voids were present at the fractured locations of the rotor cage. These defects had seriously reduced the integrity of the material and had made a major contribution to the failures. The rotor cage was made in a one-piece Al alloy casting, which is an accepted mode of construction. However, the necessary stringent NDT that is normally applied to such castings to ensure soundness seemed to have been ineffective.

FIGURE 5.21
(See Colour Insert.) Close-up view of a typical fracture appearance on the removed ER-DE, containing heavy discolouration and possible arcing damage (at the top right and bottom left), gas voids (the round and oval holes), and micro-shrinkages (all over, fir-tree appearance). The fracture mechanisms were obscured by these features.

5. The stop-start type of operation had caused two types of repeating stresses that will promote fatigue cracking. Firstly, there will be thermal stresses from differential thermal expansion and contraction in the heating up and cooling down processes when current flows on and off (twice a day). This is because the Al alloy has a thermal coefficient of expansion that is about 50% higher than that of steel. Thermal cycling of this nature causes thermal fatigue. Secondly, there will be high centrifugal stresses when the rotor is rotating at 3000 rpm and zero stresses when it is at rest. This causes a repeated mechanical stress that results in mechanical fatigue. However, the combination of the two stresses should have been anticipated at the design stage and should have been accounted for, but the design would assume sound metal and not highly defective metal, as found in this case.

FIGURE 5.22
Motor M3, removed ER-DE with the fracture surface machined away. A multitude of gas voids were present all over the machined face.

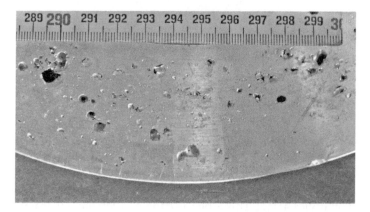

FIGURE 5.23
Magnified view of the lower-right region from Figure 5.22, showing grouped voids exceeding 10×5 mm in size.

6. The lack of condition monitoring instrumentation would not have prevented the failures but would have allowed preparations to be made to alleviate the consequences.

7. Incidentally, the third motor (M1) failed a short while later in the same manner.

5.3.4 Conclusion

1. Failure of the M2 and M3 motors had been caused by highly defective rotor bars, which had arisen out of poor manufacturing practice. These defects had not been detected after manufacture.

2. The engineering process in the selection of the motor instrumentation and in the design of the working space was not properly executed and had resulted in delays in detecting the impending failures.

5.4 Case Study 3: Damage to a Transformer Due to Water Ingress

5.4.1 Some Basic Principles Concerning Power Transformers

A transformer (Tx) is used to step up or step down voltage as required and large power transformers invariably use oil as the insulating and heat transfer medium. Newly made, degassed, and filtered transformer oils will have good dielectric strengths but when subjected to electrical stresses and heat in operation, degradation (or aging) will occur, which cause the strength to reduce. Degradation is a time-related oxidative process caused by the action of oxygen and/or other oxidants. This results in the formation of various gases and moisture, which have lower dielectric strengths than oil. Figures 5.1 and 5.2 illustrate the dangers of operating with contaminated oil. The final or near final product of oil degradation is called *sludge*, a partially conductive solid with a colour ranging from brown to black. The presence of water is particularly harmful, as it by itself will lower the dielectric strength of the oil and in addition will accelerate oil degradation. Stringent care must be taken to ensure that it does not enter into transformers through leaking gaskets or other means.

The solid insulation used in transformers is mainly cellulosic in nature and is made into various forms known as *kraft paper, boards, laminations, bars,* or *wood*. The cellulose polymers contain beta glucose units linked to each other via glycosidic bonds. They have naturally good electrical insulating properties and high dielectric strengths, which are improved by the impregnation of the transformer oil. The oil fills up naturally occurring voids in the insulation and prevents the occurrence of partial discharge across them. Cellulose insulation is degraded by three major factors, namely heat, moisture, and oxygen. Heat degradation is called *pyrolysis*, which causes cleavage of the glycosidic bonds and the formation of various compounds. Water degradation is called *hydrolysis*, which again causes cleavage of the glycosidic bonds; the action of water is catalyzed by acids (e.g., from oxidation of mineral oil). The cellulose molecule is also oxidized by oxygen (oxidation), which decomposes the glucose rings and forms water. Hence, the hydrolysis and oxidation reactions are not independent of each other. The change in the glucose structure weakens the glycosidic bonds and eventually leads to their cleavage. Degradation of cellulose reduces electrical and mechanical strengths and also causes solid sludge. These processes are slow and occur over time.

A newly made Tx will normally be subjected to a factory acceptance test (FAT) witnessed by interested parties to verify that it has been constructed in accordance with specifications. The Tx in this case study had successfully passed a FAT conducted at the manufacturer's factory. The Tx was supplied to the *Owner* by a *Vendor*, who obtained it from a reputable *Manufacturer* approved by the Owner.

5.4.2 Background Information

After the FAT, the turrets were removed (to facilitate transportation) and oil was filled to above the level of the coils; the tank was then pressurized with nitrogen to remove remaining air (and oxygen). A pressure gauge was mounted on the tank to detect any leakages that might occur through leaking gaskets. The Tx arrived at the substation in good condition with the nitrogen pressure unchanged. Further, instrumentation mounted on the tank to record acceleration forces indicated that there had not been any excessive knocks or impacts. On arrival, the vendor installed the accessories, including fitting the turrets and bushings and pumping in nitrogen, though the oil was not filtered. New gaskets supplied by the manufacturer were said to have been used and were said to be in good condition. As the site was not yet ready for commissioning, the Tx was left in this condition. About three weeks later the nitrogen pressure was found to be low and when nitrogen was pumped in, a hissing sound was detected at the Rph turret. The original turret gasket was discarded and replaced with a new one, which proved to be satisfactory. No satisfactory explanation was given on the condition of the discarded gasket and the reason why it had to be discarded after only three weeks of use under no-load conditions (*note that the life of good quality gaskets is measured in decades and not weeks*). Apparently, during gasket replacement, no examination was made of the condition of the components within the tank and no oil filtration/vacuuming was conducted to remove any water that had possibly entered. In the course of the investigation, the meteorological records were examined and it was found that during the three-week period, there were several occasions when heavy rain fell.

No further action was taken for more than a year, except for periodic examination of the nitrogen pressure until the Tx was ready to be commissioned. The oil was then subjected to hot filtration and vacuuming. Tests before filtration showed the dielectric strength to be below specifications but it improved to acceptable levels after. However, no moisture tests were conducted before filtration. Electrical tests were then conducted by the owner, which showed that the SFRA (sweep frequency response analysis) and

FIGURE 5.24
View of a similar Tx (transformer) at the substation; LVCB is the low voltage cable box, OCT is the oil conservator tank, and MT is the main tank containing the transformer coils. The LVCB end is designated as the 'fore' end, as per marine technology.

Megger (megaohm meter) results were unsatisfactory. Unsatisfactory SFRA results indicate changes in the relative positions of the major components within the tank (see Problem 5.6) and low Megger results show that the electrical insulation had been compromised. Re-testing after two more filtration exercises did not give acceptable results and the decision was made to send the Tx back to the production factory for teardown and examination. Before that, the FI made a familiarization visit to the substation to inspect a similar model of Tx (Figure 5.24).

5.4.3 Physical Inspection at the Production Factory

The teardown examination was witnessed by the factory personnel, the vendor, the insurance adjuster, and the FI commissioned by the insurance company. Initial examination and electrical testing was conducted after removal of the oil but before removal of the tank cover. Figure 5.25 shows a bird's eye view of the Tx at the factory, with the fore and aft ends indicated. The locations of the opened up HV turret ports at the Rph, Yph, and Bph are also indicated. The leaking gasket occurred at the Rph turret port. Inspection of the nameplate information indicated that it had a power output

FIGURE 5.25
Bird's eye view of the Tx at the manufacturer's factory, with the fore and aft ends indicated. The locations of the opened up HV turret ports at the Rph, Yph, and Bph are also indicated. The leaking gasket occurred at the Rph turret port.

of 60,000/90,000 kVA and a step down ratio of 132/33 kV. Figure 5.26 shows the main tank, viewed from the port side; the turrets of the HV ports of all the three phases had been removed and the covers opened up to permit electrical testing. A view of the inside of the tank through the Rph turret port revealed the presence of a darkish colouration on the top surfaces of the insulation boards (Figure 5.27). A view of an adjacent location showed that the top edges of the laminations of the steel core contained reddish rust and that the top surfaces of the end insulator boards had a dark colour (Figure 5.28).

A further examination was conducted after removal of the tank cover (Figure 5.29). The horizontal surfaces of the insulation boards at the Rph windings could then be seen to contain large areas of partially dried, darkish-coloured sludge (Figure 5.30). The steel core at this region contained large but localized areas of reddish rust at the top of the laminations (Figure 5.31). The floor of the tank at the base of the Rph winding, when viewed from the aft end, was seen to contain a fairly large area with old water stains and darkish blobs of sludge (Figure 5.32). A vertical, rectangular support bar was present, which stretched to the top of the winding and helped to secure the parts together. The same area, when viewed from the port end, showed that the base of the support bar had darkened, swelled, and split along its length and that the insulating laminations supporting the winding had warped (Figure 5.33).

FIGURE 5.26
Top of the main tank, viewed from the port side: The turrets of the HV ports of all three phases have been removed and the covers opened up to permit electrical testing. The Rph port is closest to the lower end of the figure.

FIGURE 5.27
(See Colour Insert.) View of the inside of the tank from the Rph turret port; arrows point to darkish colouration on the top surfaces of the insulation boards just below the port.

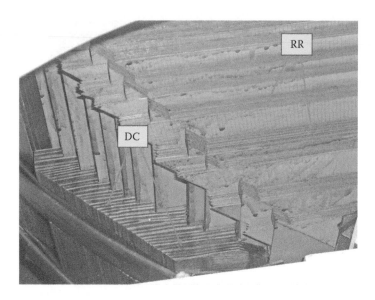

FIGURE 5.28
(See Colour Insert.) View of the inside of the tank a little further from Figure 5.27, showing the top edges of the laminations of the steel core to contain reddish rust (RR) and the top surfaces of the end insulator boards to have a dark colour (DC).

FIGURE 5.29
The Tx cover being lifted up after removing the weld line at the base of the steel tank.

FIGURE 5.30
(See Colour Insert.) After removing the tank cover, the top of the Rph winding is seen to have large areas of partially dried, darkish-coloured sludge on the horizontal surfaces of the insulation boards.

FIGURE 5.31
(See Colour Insert.) The steel core laminations below the Rph turret port contained large but localized areas of reddish rust at the top of the laminations.

FIGURE 5.32
(See Colour Insert.) The floor of the tank at the base of the Rph winding contained a fairly large area with old water stains and darkish blobs of sludge (seen from the aft end). A vertical, rectangular support bar was present, which stretched to the top of the winding and helped to secure the parts together.

FIGURE 5.33
(See Colour Insert.) The same area as shown in Figure 5.32, but viewed from the port end; the base of the support bar had darkened, swelled, and split along its length. Further, the insulating laminations supporting the winding had warped.

5.4.4 Laboratory Testing

During teardown inspection at the factory various electrical tests were conducted, which showed that the electrical components had been compromised and needed to undergo major repairs if it were to attain its original specifications. Further, various samples were acquired for further testing. They were:

1. Oil samples at various locations (performed at factory).
2. Samples of the dark sludge on the surface of the insulation boards (performed at a third-party laboratory).
3. Samples of visibly unaffected support bar from the top of the bar shown in Figure 5.33 (performed by FI).

Results from item 1 showed that the moisture levels and dielectric strengths of the samples were below acceptance levels. Results from item 2 (using SEM/EDS) showed a predominance of carbon and oxygen, deemed to be oxidation products of cellulose. For item 3, bar samples were soaked for 23 days in closed bottles containing pure oil, a water/oil mixture, and pure distilled water (Figure 5.34). An examination of the samples after testing showed that where they were in contact with water, they had swelled, cracked, and darkened slightly (Figure 5.35).

5.4.5 Analysis of Results and Conclusions

1. A leaking gasket was present at the Rph HV bushing port of the Tx for 23 days before discovery. Meteorological records showed that

FIGURE 5.34
Laboratory tests with sections cut out from the bar are shown in Figure 5.33. The left bottle is completely filled with oil; the middle bottle has water and oil; and the right bottle has only water. 'OL' and 'WL' stand for 'oil level' and 'water level', respectively.

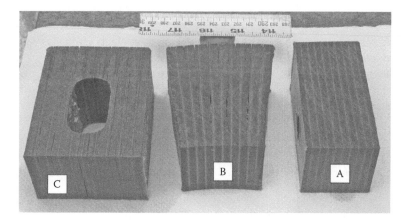

FIGURE 5.35
After 23 days of testing, Sample A remained unchanged; Sample B had swelled and cracked at its bottom; and Sample C had swelled and cracked over its length. The bottom of the samples faces the top of the figure.

during this period the site experienced some days of heavy rainfall. Large quantities of water and oxygen were able to enter the Tx tank, where they remained for more than a year before electrical testing was done by the owner. When the electrical tests repeatedly failed, the Tx was subjected to teardown examination.

2. An examination showed that the cellulose insulation and support components that were in contact with the water had become irreparably degraded; hydrolysis and oxidation were the primary degradation mechanisms. The process was slow and sustained and had occurred over the approximately one-year period when water and oxygen first entered, up to the time when filtration was carried out prior to the attempt to commission. Affected insulation would have to be replaced.

3. The same mechanism as in item 2 above would also have caused degradation of the cellulose paper insulation of the windings, and in addition, conductive particles could have become entrapped within the paper. These considerations resulted in all three windings being declared by the manufacturer and owner to be in need of refurbishment; the FI concurred with this decision.

4. The oil insulation was less affected and could still be rehabilitated by proper treatment.

5. The presence of water and oxygen had also caused the steel laminations of the transformer core at certain locations to become corroded and unusable in the discovered condition.

6. The vendor had erred by not having carried out any inspection/testing to determine whether any rainwater had entered when the

leaking gasket was initially discovered. If the presence of water had been discovered at that time and filtration performed, the damage could have been prevented or reduced.

7. There was no evidence of any lack of quality in the Tx or its accessories.

8. There was also no evidence to suggest that water had been introduced deliberately.

Problems and Answers

Problem 5.1

A transmission line, 32 kV and 50 km in length carries a current of 500 A using an Al conductor. The resistivity of Al can be assumed to be 3.2×10^{-8} Ω.m at operating temperatures. Assuming that the current is evenly distributed within the conductor strands, calculate the following:

(a) The required conductor area to restrict heat loss to 25 W over the length of conductor.

(b) If the voltage were to be changed to 132 kV using the same conductor, what would be the heat loss?

(Note that this problem only serves to illustrate some principles and that design practice needs to consider the various standards and codes prevailing.)

Answer 5.1

(a) Using Ohm's law: $H = I^2R$, $R = (25)/(500)^2 = 100.0 \times 10^{-6}$ Ω

Area required, $A = (3.2 \times 10^{-8})/(100.0 \times 10^{-6}) = 320.0 \times 10^{-6} \times 1000^2 = 320$ mm^2

(b) Using $V = I \times R$, at constant R and V2 = 132 kV, $I_2 = (500)(32/132) = 121.2A$.

The heat loss = $25 \times (121.2/500)^2 = 1.47$ W

This illustrates the advantage in using high-transmission voltages.

Problem 5.2

How can you detect the following by portable equipment: (a) excessive ohmic heat in bolted joints; (b) partial discharge on the surface of a transformer bushing.

Answer 5.2

(a) Thermographic equipment is sensitive to infrared or UV light; infra-red is good for low light at night and UV is good for daylight.

(b) Acoustic sensors that can distinguish the ultrasonic waves made by PD from the background noise.

Problem 5.3

What are electrical 'trees?'

Answer 5.3

The aptly named *trees* are damage marks in the interior of polymeric insula-tors made by internal partial discharge. They look like trees when the affected insulator is sliced up, stained, and examined under an optical microscope. Certain polymers are more resistant to treeing than others, for example the TR XLPE (tree resistant cross-linked polyethylene).

Problem 5.4

What is the Duval triangle that is used in analyzing the condition of OLTCs?

Answer 5.4

During operation, OLTCs continuously make and break low currents, which nevertheless cause breakdown and degradation of the insulating oil. Oil breakdown results in the evolution of gases and the relative amounts of three gases, methane (CH_4), ethane (C_2H_4), and acetylene (C_2H_2) are used in the Duval triangle to predict the events occurring within the OLTC, and ultimately, its condition. The basis for the triangle is that events of different energy levels will occur at different temperatures, which will evolve different types of gases. The gases CH_4, C_2H_4, and C_2H_2 are formed in order of increas-ing temperatures. If the three gases are plotted on the triangle, their relative amounts can usefully predict the presence of conditions such as low-energy DD, high-energy DD, and low-, medium-, and high-range thermal faults.

Problem 5.5

Rotor bars in large induction motors are often made of a one-piece cast Al alloy. (a) What are the pros and cons of such manufacture? (b) What steps should be taken to ensure that the bars will not fail in service?

Answer 5.5

(a) If a fabrication process were to be used, the rotor bars (say, 39 off) need to be accurately machined and closely fitted into slots in the steel core, and then joined to the end ring by welding; this would require precise and laborious work. Casting molten metal into the formed slots and allowing them to fill up the space is a neat solution. However, Al castings are prone to gas voids if the molten metal is not properly degassed and also to trapped air if the casting is of complex shape and not properly vented. Further, if the casting process is not designed to allow for directional solidification, micro-shrinkages will occur. Both these defects were present in large quantities in M2 and M3.

(b) Assuming that the material has been chosen correctly, its integrity must be ensured after it has solidified. For defects between the ends of the bars and the end ring, simple ultrasonic testing will suffice. For large internal defects and breakages within the bars, an electrical test known as a *growler* test may be effective (do your own reading on a growler test).

Problem 5.6

How does SFRA work?

Answer 5.6

From Wikipedia, Sweep Frequency Analysis at http://en.wikipedia.org/wiki/Sweep_Frequency_Response_Analysis:

Sweep Frequency Response Analysis (SFRA) is a powerful and sensitive method to evaluate the mechanical integrity of core, windings and clamping structures within power transformers by measuring their electrical transfer functions over a wide frequency range. SFRA is a proven method for frequency measurements.

Transformers consist of multiple complex networks of capacitances and resistors that can generate a unique signature when tested at discrete frequencies and plotted as a curve. The distance between conductors of the transformer forms a capacitance. Any movement of the conductors or windings will change this capacitance. This capacitance being a part of complex L (inductance), R (Resistance), and C (Capacitance) network, any change in this capacitance will be reflected in the curve or signature. An initial SFRA test is carried out to obtain the signature of the transformer frequency response by injecting various discrete frequencies. This reference is then used for future comparisons. A change in winding position, degradation in the insulation, and so forth, will result in a change in capacitance or inductance thereby affecting the measured curves.

Tests are carried out periodically or during major external events like short circuits and the results are compared against the initial signature to test for any problems. The basic functionality of SFRA supports the following measuring modes. The voltage transfer function Uo/Ui (f) SFRA test reveals if the transformer's mechanical or electrical integrity has been compromised. SFRA analysis can detect problems in transformers such as:

- Winding deformation—axial and radial, like hoop buckling, tilting, spiraling displacements between high and low voltage windings
- Partial winding collapse, shorted or open turns
- Faulty grounding of core or screens; core movement
- Broken clamping structures
- Problematic internal connections

References

General References

Arora, Ravindra and Wolfgang Mosch, *High Voltage and Electrical Insulation Engineering,* Institute of Electrical and Electronics Engineers, New Jersey: John Wiley & Sons, 2011.

Grigsby, Leonard L., Ed., *Electric Power Generation, Transmission and Distribution,* 3rd ed., Boca Raton, FL: CRC Press/Taylor & Francis Group, 2012.

Stone, Greg C., Edward A. Boulter, Ian Culbert, and Hussein Dhirani, *Electrical Insulation for Rotating Machines,* New Jersey: IEEE Press, 2004.

Specific References

1. Arora, Ravindra and Wolfgang Mosch, *High Voltage and Electrical Insulation Engineering,* Institute of Electrical and Electronics Engineers, New Jersey: John Wiley & Sons, 2011, p. 369.
2. Cahier Technique No. 198, *Vacuum Switching,* Schneider Electric, 2000. (URL: www.schneider-electric.com)
3. ABB Publication 1ZSE 5491-104 en, Rev. 9, On-Load Tap-Changers, Type UZ, Technical Guide. (URL: www.abb.com/electricalcomponents)

6

Case Studies—Boilers
and Boiler Components

6.1 Introduction

Steam boilers have been used for many years in industries to run production equipment, drive steam turbines to generate electricity, provide heat to certain processes, steam-clean, and so on. A steam boiler converts water to steam with the heat energy from a fuel source such as oil, gas, and coal. The most common types of boilers are the firetube boiler and watertube boilers in which fire or hot gases are directed inside firetubes and outside watertubes, respectively. In the firetube boiler, the firetubes are arranged in banks so that the hot gases will pass through the multiple pass firetubes before they exit the stack. Water that surrounds the firetubes absorb heat from the surface of the tubes. In a watertube boiler, watertubes are arranged in a vertical position. It usually has two or more drums in which the top drum allows the separation of steam whereas the bottom drum collects sludge.

Although the technology of a steam boiler has existed for a long time, boiler system failures are still common. The failures can be catastrophic, even to the extent of causing an explosion as it is operated under high temperatures and pressure. Boiler tube failures (BTFs) are common due to the operating conditions under high temperatures and pressure. In fact, BTFs are caused by many factors involving the sources (feedwater, fuel), the processes (combustion, steam generation), and the boiler components themselves. Results of the analysis of most BTF incidents reveal that the failure could have been prevented if proper measures had been taken. Overheating and corrosion are usually the major causes of BTF.

BTF due to overheating is commonly associated with deposits. A detailed examination of the failed tube section provides an understanding whether the failure is due to conditions that cause rapid elevation in tube wall temperature or a gradual buildup of deposit. The buildup of deposits could be harmful in many ways. Among other things, deposits restrict flow causing water starvation, affect local heat transfer causing a temperature rise, and induce corrosion. Metallographic examinations are useful in understanding the overheating process and confirming whether a gradual or rapid over-heating condition existed prior to failure. In the case of rapid elevation in tube temperature to A3 temperature, plastic deformation occurs in the tube due to the loss of strength at a high temperature and rupture occurs. Rupture is identified as a 'thin-lipped' burst, a ductile fracture characterized by thin and sharp edges [1]. Further analysis of the deposits helps in determining the proximate or root cause.

In the case of BTF due to a long-duration scaling condition, the symptoms of such failure are a bulged external surface and a thick-lipped fissure [2]. This failure could be interpreted as a long-term creep failure as a result of repetitive scale formation that causes overheating and swelling (bulging) of the tube surface. The mechanism of such a failure is that the scale tends to crack off due to local temperature rise, and then water contacts the metal at the crack and cools the metal until further scaling. The processes of further scaling, crack, and water cooling at the affected region of the tube repeats over and over many times until failure occurs.

BTF due to corrosion at a high temperature can be caused by various forms of corrosion mechanisms. The corrosion mechanisms include stress corrosion cracking, caustic embrittlement, corrosion fatigue, stress-induced corrosion, dissolved oxygen, caustic attack, steam blanketing, acidic attack, corrosion due to copper diffusion, hydrogen attack, or embrittlement [3]. Contaminants and dissolved gases in the feedwater are key agents of corrosion. For instance, dissolved oxygen causes oxygen pitting on the surface of materials, while contaminants such as copper diffuse into steel and then induce cracking. Deaeration in boiler operation is an important water treatment process that eliminates contaminants and gets rid of dissolved gases, especially oxygen in the feedwater.

When conducting a boiler failure analysis, the operating conditions prior to the failure should be examined and the design specifications should be referred to. This will reveal whether the operating conditions have exceeded the design and boiler tubing materials constraints [4].

Common steel materials used in steam boiler tubes are the plain carbon steels and the steels with various amounts of alloying additions such as molybdenum and chromium, and various other grades of stainless steels suit-able for the operating temperature and capacity of the boiler concerned [4]. The material selection for a steam boiler tube is the most critical for the superheater tubes and reheater tubes. High resistance to flue gas attacks and to creep degradation are required in power plant boilers, whose design has progressed from sub-critical conditions three decades ago to the present supercritical and ultra-supercritical conditions. The latter two run at higher temperatures than sub-critical, have higher boiler efficiencies, and consequently produce lower amounts of climate-change gases (such as carbon dioxide) per unit output. Tube materials that are commonly used include

TABLE 6.1

Common Steels Used in Boilers

Materials	Materials Specification	Max. Metal Temperature (°C)
Carbon steel	ASME SA-210, BS 3059	425
Carbon ½ Mo steel	ASME SA-209 T1	470
1¼ Cr-½ Mo steel	ASME SA-213 T11	538
1 Cr-½ Mo steel	ASME SA-213 T12	538
2¼ Cr-1 Mo steel	ASME SA-213 T22	565
2¼ Cr-0.3 Mo steel	ASME SA-213 T23	595
9Cr-1Mo-½V steel	ASME SA-213 T91	650
Austenitic stainless steel	ASME SA-213 TP304, TP316, TP321, TP347	>650

carbon steel, carbon-molybdenum steel, chromium-molybdenum steel, and austenitic stainless steel. Some examples of boiler tube materials are displayed in Table 6.1; higher grades like the ASME SA-213 T92 are already in use and more exotic grades are in the pipeline. Carbon-molybdenum steels have better creep rupture strength than the carbon steels due to the molybdenum content. The chromium content in some grades of tubing materials enhances the oxidation resistance. Table 6.1 documents the maximum metal temperature for different materials (also known as the design temperature), which are based on the temperature at which graphitization sets in, or the oxidation limit at which oxidation accelerates. The actual temperatures that are used are usually lower and depend upon the steam pressure used and the creep lives required. Such a design is closely regulated by codes of practice, for example, the ASME (American Society of Mechanical Engineers) Boiler and Pressure Vessel Code.

6.2 Case Studies

This chapter presents three case studies of boiler tube failure analysis. The first case concerns an edible oil company where a boiler exploded as a result of overheating. The root cause was the malfunction of the relay or switches in the controllers that caused a water disruption. The second case involves a palm oil mill where a boiler tube burst during the preheating stage. The failure was due to water starvation, compounded by cracking from intergranular copper penetration. The third case is an incident of boiler furnace tube failure in a power plant where furnace tubes ruptured because of creep cavitation. The root cause was the formation of oxide layers on the tube surface that led to overheating and plastic deformation.

6.3 Case Study 1: Explosion of a Boiler in an Edible Oil Company

In an edible oil company, a firetube boiler exploded and caused two fatalities. The boiler was a support boiler and only came online when the main biomass-fired boiler could not produce sufficient steam. A forensic investigation was called for to identify the root causes of the incident, focusing mainly on the nature of the failure and the physical conditions that led to the blast. Metallurgical investigations were initially conducted on the furnace tubes in the as-received condition and subsequently they were subjected to various heat treatments to simulate service conditions. Results of the analysis revealed that the top one-third of the furnace tube overheated to nearly 700°C; the material had lost much of its strength and modulus of elasticity at such a high temperature, causing it to collapse under normal, external water pressure. The overheating was caused by an aggravated low-water condition. Due to the malfunction of the relays/switches in the controllers, the water level dropped continuously and passed the L1 (first low-water level) and L2 (second low-water level) without activating the controllers.

6.3.1 Background Information

Figure 6.1a shows a photograph of the boiler before it exploded. The starboard side (located at the direction of 3 o'clock) was not visible, but contained a second water gauge and two Mobrey water controllers. The first controller (M-L1) is a modulating cum low-water level/burner cutoff controller, and the second controller (H-L2) is a high cum extra low-water level/burner lockout controller.

Note that M-L1 has a dual function, that is, a modulating function and a low-water alarm/fuel cutoff function. The controller allows the burner to reignite if the water level rises to normal. H-L2 is a separate system from M-L1 and works independently by itself. The controllers function via magnetic switches in the head, which will operate when a magnet on the float spindle passes over them. There is one switch for high water and one for low water. M-L1 and H-L2 together form a very safe combination if all the key components are working properly. The nameplate on the boiler indicates the following specifications:

a. Type: Firetube boiler
b. Rated capacity: 11,360 kg/hr from 100°C
c. Design pressure: 1900 kPa
d. Working pressure: 1900 kPa
e. Hydrotest pressure: 2850 kPa.

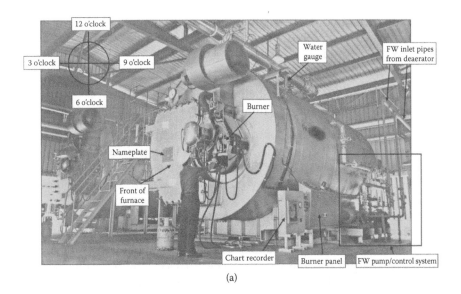

(a)

(b)

FIGURE 6.1

(a) The boiler when it was new; the front of the boiler is termed 'fore' and the rear is 'aft'. The clock face shows the relative transverse locations; the 'port' side is at 9 o'clock and the 'starboard' side at 3 o'clock. (b) The top of the boiler with all the steam fittings broken off; an ellipse encloses remnants of the two Mobrey controllers.

6.3.2 Preliminary Considerations

Information from the pressure chart shows that the boiler was operating from approximately 7:30 P.M. to 9:20 P.M., when the explosion was thought to have occurred. The time difference of about 1 hour 50 minutes is assumed to be the period during which the boiler was producing at full load. The pressure chart shows that the explosion had occurred under normal pressure.

This contradicts the hypothesis that the explosion could have been due to a sudden increase in pressure caused by water entering into the 'dry' boiler and sudden evaporation.

6.3.3 Damage to the Boiler's Shell and Furnace Tubes, and the Implications

After the explosion, the shell of the boiler flew some distance away and the boiler was lying on its port side, with all the fittings damaged or missing. Figure 6.1b shows the top of the boiler with all the steam fittings broken off and remnants of the two Mobrey controllers partially buried in the sand and mud.

An examination at a later date showed that the water inlets to the Mobrey controllers were not blocked by sludge (Figure 6.2), suggesting that the float mechanism could have functioned as intended, should there have been low water just prior to the incident. The furnace tube had collapsed and the furnace wall had separated from the tubeplate on the port side, leaving a large access gap into the second pass chamber (Figure 6.3). The collapse had occurred because the furnace was pinched horizontally on both horizontal sides. This is indicated as the 'collapse direction' in Figure 6.4. The tubeplate and reversal chamber wrapper plate had darkened significantly above the level L-1-3 (the water level at a distance of one-third diameter from the top of the furnace); the darkening was caused by the presence of fairly thick high-temperature oxides. It can be seen in Figure 6.5 (the port fore side viewed from the aft end) that about the top one-third of the external furnace tube surface, at the locations to the right of the demarcation line D1–D2, showed similar darkening. The darkening was the result of overheating to temperatures in excess of normal design. The distinct difference in appearance between the darkened and non-darkened regions points to a pronounced temperature difference between the two regions. This could only have occurred if the non-darkened region was water-touched (flooded) and the darkened region was steam-touched; this was in fact an aggravated low-water situation. This explains how the explosion had occurred, when

FIGURE 6.2
The water inlets to the Mobrey controllers were not blocked by sludge.

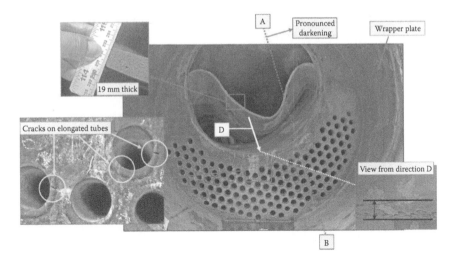

FIGURE 6.3
(See Colour Insert.) Boiler conditions at the aft end. (i) The tubeplate and reversal chamber wrapper plate show significant darkening above the horizontal AB, about one-third of the furnace diameter distance from the furnace top. (ii) The ruptured edge of the furnace wall was measured to be about 19 mm thick (inset). (iii) The ends of some second pass tubes had become oval and developed cracks, (iv) when viewed in direction D, the part of the tubeplate close to the furnace tube can be seen to have been pulled inward towards the tube, by deformation of the tube during the incident.

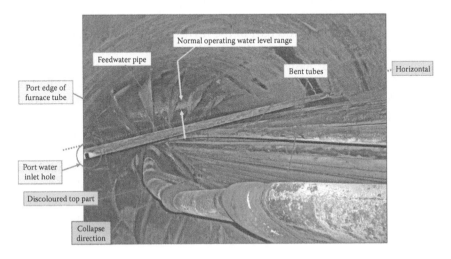

FIGURE 6.4
(See Colour Insert.) View of the outside of the furnace tubes, as seen from the aft end: The first pass furnace tube had collapsed by being pinched horizontally on both sides. Some of the second pass tubes had been slightly bent. The top one-third of the furnace tube surface had a much darker colouration than the remainder of the shell. The port water inlet hole was located beyond the port edge of the furnace.

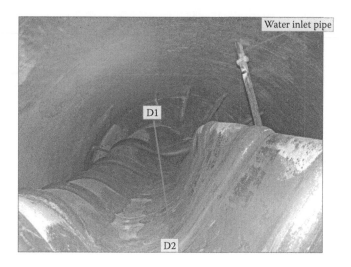

FIGURE 6.5
(See Colour Insert.) View of the port, fore side of the (first pass) furnace tube, from the aft end the top one-third of the tube surface is darkened, with a sharp demarcation between the darkened and non-darkened parts.

the water was at, or was just below the lower part of the dark zone, at L-1-3. Since time is required for the furnace wall metal to become overheated (from the normal temperature of less than 300°C to about 700°C), and also for the dark oxides to form, the situation must have persisted for some time, likely for 20 minutes or more.

The fact that the entire top part of the furnace tube and the tubeplate had been overheated shows that overheating had not been due to burner misalignment, which normally only affects a small zone near the burner. The second and third pass tubes (from the bottom of the furnace tube) had not been darkened (see sample F34-6 in Figure 6.6), showing that their temperatures had not reached that of the dark zone of the furnace. It can be seen clearly in Figure 6.4 that the normal operating water level was within a small range, and was well above the level of the topmost tubes and feedwater inlet pipe. The small operating water level range indicates that the modulating control valve was working normally. There was no evidence that inadequate material, welding, or design had contributed to the failure. The overriding factor was that it had been heated to about 700°C, at a temperature which its strength would have been much lower than usual.

6.3.4 Damage to the Boiler Fittings, and the Implications

The electrical wiring and components in the burner panel were smashed up and they could not be examined at all. It was quite impossible to make

FIGURE 6.6
(See Colour Insert.) Furnace samples F34-6 and F34-12 from between bowling rings 3 and 4; the sample F34-12 from the top of the tube was darkened but F34-6 was not.

anything out of the mess. The control box for the modulating valve was also damaged and most of the wires and the control relays were torn off. It is most unfortunate that the relays were not available for examination as they play a key role in the controlling function of the control box. The welding up of the electrical contacts or mechanical sticking would cause them not to operate when required to do so at low water.

The float system of both M-L1 and H-L2 appears to be in good condition, but the head and internal components were missing so they were unavailable for examination. The modulating control valve was found with its spindle pin stuck in the 40% open position (Figure 6.7). When the cylinder was opened up, large amounts of corrosion products and dried grease were found (Figure 6.8a). After removing the cylinder, it can be seen that the spring had corroded and thinned down significantly (Figure 6.8b) indicating that the control valve had not been cleaned for the inspection.

Figure 6.9 is an illustration of the valve spindle in a fully opened position. The bottom half of the valve spindle looked burnished due to constant operation but the top half is rusty. The top level of the burnished part (L) roughly corresponds to the pin position shown in Figure 6.7. A small length of the rusty part, LB, is only lightly burnished (inset in Figure 6.9). It is obvious that the spindle had not been pushed below the burnished level for a long time, otherwise its whole length would have been burnished. This shows that the valve does not normally close more than about halfway down. In the event that the spindle is forced beyond the burnished portion, into the rusty part, it would quite likely get stuck there and not be returned by the rusty and weakened spring (Figure 6.8b).

FIGURE 6.7
The VH2 modulating control valve: the pin position shows the valve to be stuck in a 40% open position. The lever connecting to the actuator of coil B was bent towards the 'close' position.

(a)   (b)

FIGURE 6.8
(a) The top of the cylinder after the cover was removed; large amounts of corrosion products and dried grease were present. The water normally present in the cylinder had leaked out. The spindle nut needs to be unscrewed before the piston plate can be removed. The piston is kept watertight by the piston cup. (b) After removing the cylinder: The spring was heavily corroded even though the chamber was supposed to be dry.

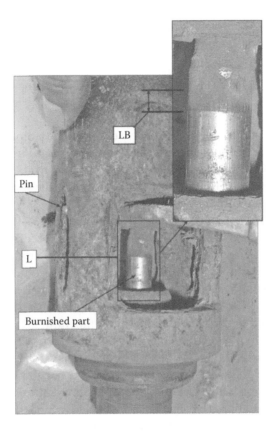

FIGURE 6.9
The valve spindle in a fully opened position: the bottom half of the valve spindle has been burnished due to constant operation but the top half was rusty. The top level of the burnished part (L) roughly corresponds to the pin position shown in Figure 6.7. A small length of the rusty part, LB, was very lightly burnished (inset).

6.3.5 Results of the Laboratory Examination, and the Implications

Selections of the most viable samples are first carried out. The samples initially examined were named as:

a. F34-6: First pass furnace tube between bowling rings 3 and 4, at the 6 o'clock position (specimen F34-6).

b. F34-12: First pass furnace tube between bowling rings 3 and 4, at the 12 o'clock position (specimen F34-12).

c. 2T5: Second pass tube, both the rear and front ends (specimens 2T5).

Additional specimens were cut out from Sample F34-6, which were normalized at 900°C and then tempered at 600° and 700°C, respectively. These specimens are labeled as:

  a. Specimen F34-6-N9: Normalized at 900°C, not tempered.

  b. Specimen F34-6-N9-T6: Normalized at 900°C, tempered at 600°C for 1.5 h.

  c. Specimen F34-6-N9-T7: Normalized at 900°C, tempered at 700°C for 1.0 h.

The main aim of the heat treatment is to establish the temperatures at which they had been heated to, which can be estimated from the level of thermal degradation suffered by the pearlitic component of the microstructure. Keep in mind, however, that the furnace tube (first pass) had been stress-relieved after welding, to temperatures between 580° and 620°C.

The top part of the first pass furnace tube (Sample F34-12, see Figure 6.6) had been heated up to nearly 700°C before the failure occurred. This is much in excess of the design temperature of 262°C for the front and rear tube-plate, which is presumably higher than for the furnace tube. Figures 6.10a (magnified 200×) and 6.10b (magnified 1000×) show visible degradation of the pearlite [5] for F34-12 at 3 mm from the inner surface, indicating that it had been exposed to a significantly higher temperature during its lifespan. However, the bottom part suffered only about 600°C, possibly from the stress relief applied after welding. No definite conclusions can be made here.

The samples of the second pass tubes, 2T5, did not show visible signs of pearlite degradation, which will only occur at temperatures above 550°C for short exposure duration. They had not been heated to more than 550°C, meaning they were not stress-relieved. Figure 6.11a displays specimen 2T5 which is magnified 200× at 3 mm from the inner surface: showing about 15% pearlite in a matrix of ferrite. Specimen 2T5 was magnified at 1000× (Figure 6.11b) at 3 mm from the inner surface, showing no visible thermal degradation of the pearlite. Unidentified precipitates are also present. Similar observations can be seen for the other sample of the second pass furnace tube.

6.3.6 Most Likely Scenario of the Failure

From the results, the following failure scenario was proposed as being the most likely to have occurred:

  a. During the steam demand period after the main boiler was taken offline, the modulating valve got stuck at a low position and the water intake could not keep up with steam delivery.

  b. The water level dropped continuously, and water passed the L1 and L2 levels without activating the controllers, the reason being electrical or mechanical malfunction in both controllers (specifically the relays/switch).

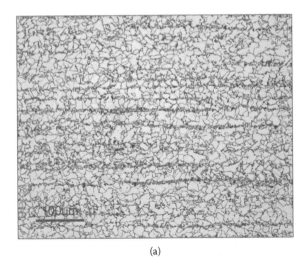

(a)

(b)

FIGURE 6.10
(a) Specimen F34-12: 3 mm from the inner surface, showing visible degradation of the pearlite. Mag. 200×. (b) Magnified view, showing that thermal degradation of the pearlite was quite pronounced. Mag. 1000×.

    c. The top one-third of the furnace tube had been overheated to nearly 700°C causing the material to lose much of its strength and modulus of elasticity, and eventually it collapsed under normal pressure.

6.3.7 Conclusion

The explosion was caused by a collapse of the furnace tube from external pressure. The pressure was normal but the top part of the tube was seriously

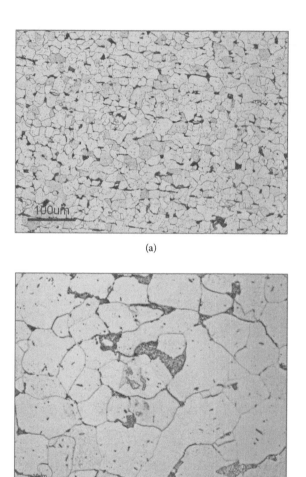

(a)

(b)

FIGURE 6.11
(a) Specimen 2T5-R-12: 3 mm from the inner surface, showing about 15% pearlite in a matrix of ferrite. Mag. 200×. (b) Specimen 2T5: Magnified view showing no visible thermal degradation of the pearlite. Unidentified precipitates were also present. Mag. 1000×.

weakened from overheating to about 700°C. The overheating was caused by an aggravated low-water condition. The possibility of burner misalignment was considered to be unlikely; so was the likelihood of a sudden pressure rise from feedwater entering into a 'dry' boiler. The low-water condition had been caused by some disruption in the feedwater supply, together with a simultaneous failure of both the Mobrey controllers. The disruption in feedwater was likely due to a bad modulating valve but this is a non-critical issue. The failures of the Mobrey controllers to cut off the burner when low water occurred were due to some malfunction of their relays/switch, and these are very critical issues.

The malfunction of the relays/switch occurred because they had not been serviced since commission. This arose from a lack of awareness that servicing was required, as no such recommendations had been made in the equipment manuals. Human error was not a factor, except for the non-detection of the low level in the visually visible water gauge. It is believed that based on the findings of this report, the relevant authorities decreed that such relays and switches must thereafter be inspected regularly.

6.4 Case Study 2: Boiler Tube Failure at an Oil Palm Mill

A boiler tube at an oil palm mill was reported to have ruptured during the preheating stage prior to full operation. Metallographic analysis and laboratory examinations were performed to investigate the reason for the failure. The tube had burst during a period of gross overheating to at least 830°C. Bursting had been facilitated by the presence of deep pre-existing cracks, caused by intergranular copper (Cu) penetration of an unusual nature [6]. Nevertheless, even in the absence of the Cu penetration and cracking, it is very likely that the tube would have eventually burst from the severe overheating. The only possible cause of overheating is water starvation.

6.4.1 Macro-Examination and Discussion

The front end of the furnace through which fuel is fed is defined as the 'fore' end, and the rear end is defined as the 'aft' end. Figure 6.12 shows the fore-end view of the water boiler; the starboard side is on the left of the figure, and the port side on the right. Arrow 'E' is pointing to the openings for raking the fire and for entry into the furnace. Inset 'A' shows the steam pressure gauges; the gauge on the left is marked 20 kg/cm^2 which indicates the maximum allowable pressure. Inset 'B' shows the nameplate bearing the following information:

a. Type: Watertube boiler
b. Max allowable working pressure: 20 kg/cm^2
c. Hydrostatic test at 33.5 kg/cm^2 on 2/11/1982
d. Year built: 1982

Other information about the tube obtained verbally from the company:

a. Tube material: BS 3059
b. Tube dimensions: 63.5 mm OD × 3.6 mm thick

FIGURE 6.12
The front (fore)-end view of the water boiler; the starboard side is on the left side of the figure, and the port side is on the right.

Figure 6.13a shows the fore side of the furnace, as viewed from the inside. The bottom row of openings is for raking and entry. The top row of openings is for feeding the fuel fibre. The burst happened at the second tube from the starboard side, one-third of the way from the top end of the front firebrick wall (Figure 6.13b, indicated by a circle). The front wall tubes had sagged, above the level of the burst (Figure 6.14a). The port and starboard side tubes were distorted and had moved from their original positions (Figure 6.14b).

Figure 6.15a,b are close-up views of the burst, with the following features:

a. A fishmouth appearance with jagged fracture paths that did not run along a single line.
b. Fairly thick fractured lips.

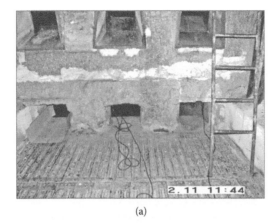

(a)

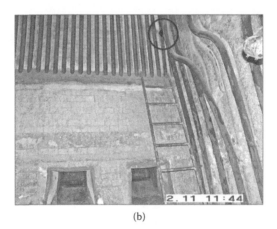

(b)

FIGURE 6.13
(a) Front (fore) end of the furnace, as viewed from the inside. (b) The location of the burst tube (circled) at the second tube from the starboard side.

   c. Two small tears near the lower tip of the main tear.

   d. The internal surface was quite free of scaling and deposits.

6.4.2 Laboratory Examination

The front wall tubes are numbered from the starboard side, beginning with Tube 1 to Tube 29 at the port side. Several samples were submitted for laboratory examination. They comprised the burst tube as well as tubes from various other locations to serve as a comparison: (i) one length of Tube 1, (ii) two lengths of Tube 2 some distance from the burst, (iii) one length containing the burst, (iv) one length each of Tubes 14, 15, 16, 28, and 29. The measurements of the supplied tube samples did not reveal any significant thinning.

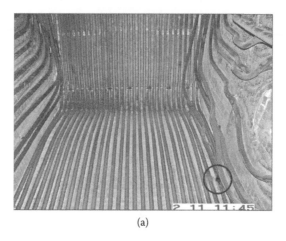

(a)

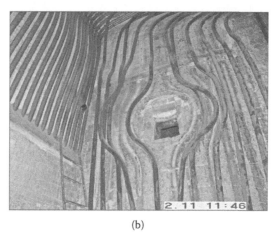

(b)

FIGURE 6.14
(a) Shows the location of the tube-burst relative to the furnace front wall and roof tubes. The port and starboard wall tubes have been displaced from their original positions. The front wall tubes appear to have sagged. (b) Shows the starboard wall tubes distorted and displaced from position. Some front wall tubes are also seen to have sagged, above the level of the burst.

6.4.3 Metallographic Examination

Figures 6.16 and 6.17 show the specimens that were cut out from the tube samples for metallographic examination. Note the following details:

a. The specimens taken at the burst are marked B, B1, and B2. In the vicinity of B2, numerous secondary cracks were present on the inside surface, parallel to the primary crack path.

b. Specimen X1 is located at the fireside from the end of the tube sample, 60 cm upstream of the burst.

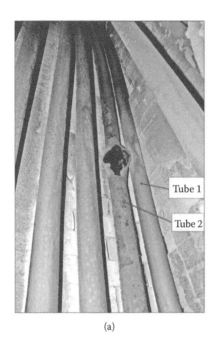
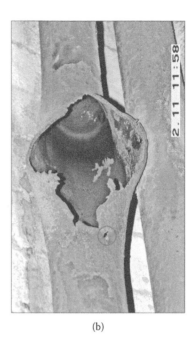

(a)   (b)

FIGURE 6.15
(See Colour Insert.) (a) Shows the burst located at the second tube from the starboard side (Tube 2). The parts of the burst tube from about 2 diameters away from the burst did not any show any visual signs of bulging. The surfaces of Tube 2, far away from the burst, and of the adjacent tubes, were covered with a thick, darkish magnetite scale. (b) Shows a close-up view of the burst.

A metallographic examination via optical microscopy was conducted on the specimens above to identify the metallurgical differences between the burst area and the areas that were remote from the burst. The basic characteristics of specimens B, B1, and B2 were fairly similar, so reference shall only be made to specimen B1. The following four features were observed:

1. Details of the fracture profile (Figures 6.18, 6.19, specimen B1):
    a. Figure 6.18a shows that the fracture profile was made up of two regions of differing characteristics marked 'PC' and 'FF'. PC started from the inside surface and took up approximately 85% of the total thickness while FF stretched between PC and the outside surface. The region PC contained pre-existing primary cracks of an intergranular nature. It was devoid of plastic deformation and the cracks were present before the final fracture occurred. Secondary cracks were present at this region and also of an intergranular nature, branching off from the primary crack (Figure 6.18b). The region FF was an area of shear fracture, with characteristics of prior plastic deformation; this was the final fracture region.

(a)

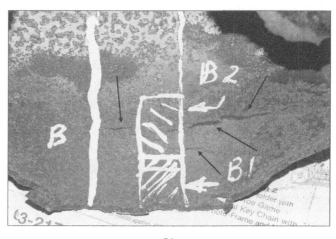

(b)

FIGURE 6.16

(a) A close-up view of the burst tube, showing the locations of specimens B, B1, and B2. The arrow points to the top of the furnace. (b) Shows a close-up view of specimens B1 and B2. The arrows are pointing to numerous secondary cracks that were present on the inside surface, parallel to the primary crack path.

b. The microstructure generally consisted of small amounts of a coarse Widmanstatten ferrite phase (light coloured) with large amounts of some other darkish phases, comprising fine pearlite and some mixed ferrite/carbide constituents (Figure 6.18a,b).

c. Magnified views of the secondary cracks show the presence of a copper (Cu) phase in the uncracked grain boundaries (Figure 6.19). The primary and secondary cracks at this location were seen to be intergranular in nature. The darkish phase was seen to be predominantly very fine pearlite, possibly with some mixed ferrite/carbide phases.

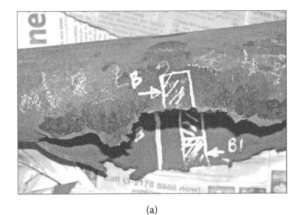

(a)

(b)

FIGURE 6.17
(a) Close-up view of specimen B. The fracture profile was seen to be jagged in nature, indicative of some form of material embrittlement. (b) Shows the location of specimen X1 on the burst tube, at the fireside face, 60 cm upstream of the burst. Another micro-specimen, X2, was taken at the same plane on the opposite side.

2. Details of the internal surface adjacent to the fracture profile (Figure 6.20, specimen B1):

   a. Fine secondary cracks were seen on the internal surface parallel to the primary crack path, similar to the coarse ones shown in Figure 6.16b.

   b. These cracks were all intergranular in nature and mostly contained a Cu phase in their uncracked grain boundaries, similar to the secondary cracks branching off from the primary crack path (Figure 6.20).

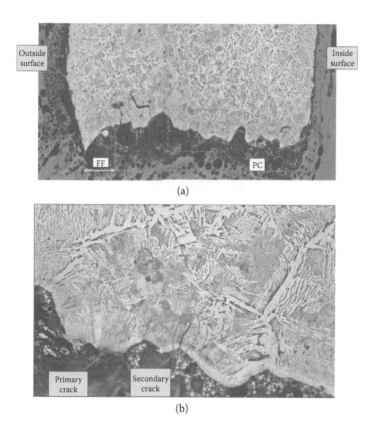

(a)

(b)

FIGURE 6.18

(See Colour Insert.) (a) Shows the fracture profile of specimen B1; the fracture profile is made up of the regions 'PC' and 'FF'. Mag. 60×. (b) Magnified view of the centre of the fracture profile (a). The primary crack and secondary crack at this location were seen to be intergranular in nature. The darkish phase was seen to be predominantly very fine pearlite, possibly with some mixed ferrite/carbide phases. The microstructure was coarse. Mag. 300×.

    c. The microstructure here comprised the same phases as seen in the fracture profile, except that the percentage of the ferrite phase was larger.

3. Details of specimen S29-2

Tube 29 did not contain any surface cracks; the microstructure consisted of medium/fine, block-shaped, equiaxed ferrite with about 20 to 25% of pearlite, equivalent to about 0.15 to 0.20% carbon. This microstructure had been formed by slow cooling rates.

4. Details of specimen X1 (Figure 6.21)

No cracks appeared in this specimen. The microstructure was basically similar to specimen B1, except for the larger ferrite percentage.

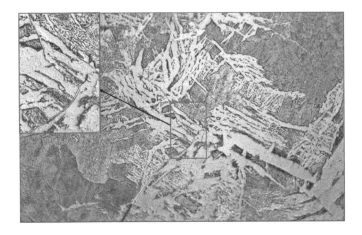

FIGURE 6.19
(See Colour Insert.) Magnified view of the secondary crack from Figure 6.18b, showing the presence of a light-coloured, copper phase in the uncracked grain boundaries above the secondary crack. Mag. 750×.

6.4.3.1 Analysis of Microstructures

The results of the microstructure analysis revealed that all the microstructures seen in specimen B1 had been formed by cooling fast from above the A3 temperature, which for 0.2% carbon steel is about 830°C. The amount of the Widmanstatten ferrite present in the microstructure generally varies with the cooling rate for the situation at hand; the faster the cooling, the lower the quantity. From this, it may be seen that specimen B1 at the fracture faces had cooled relatively much faster than at locations away from the fracture surface. This shows that the microstructures had been formed just after the burst, due to the fast exit of pressurized water/steam. The steel had not been heated up again after the burst, to temperatures beyond about 700°C, or they would have re-transformed.

The microstructure in specimen S29-2 could have been original or could have been formed at the same time as those at B1. It is only presented to show what the normal microstructure for this steel should be.

The intergranular Cu phase in specimen B1 had caused the steel to become brittle and to crack easily, even under low stress. These cracks initiated at the inside surface of the tube and had extended to about 85% of the metal thickness before the final fracture occurred. The presence of the Cu is an abnormal event; its presence could possibly have arisen in two ways:

    a. First, Cu-rich deposits had formed on the inside surface of the tube, and diffused into the steel when gross overheating occurred. This requires that some Cu components that were present or had been present in the system, had become dissolved to give rise to Cu ions,

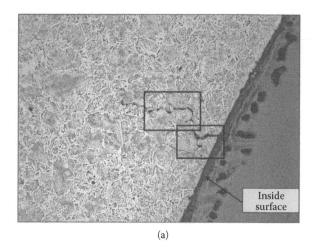

(a)

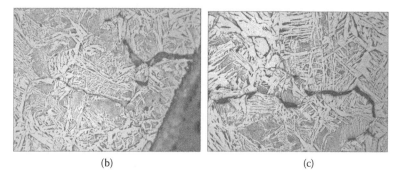

(b)   (c)

FIGURE 6.20
(a) Specimen B1: Shows the presence of a secondary crack on the inside surface, parallel to the primary crack and about 3 mm from it. Mag 50×. (b) Magnified view of the start of the secondary crack, showing the crack path to be intergranular. A copper phase is seen at some of the uncracked grain boundaries. Mag. 100×. (c) Magnified view of the area within the top rectangle in (a), showing intergranular cracking and copper at the uncracked grain boundaries. Mag. 200×.

which then deposited on the steel surface. The deposits could possibly also have come from foreign contaminants. This Cu-deposit scenario appears to be quite likely, but whether the Cu had diffused in, on this occasion or on previous occasions of overheating, is not known.

b. Second, the copper probably originated from a defective steel tube that contained copper. However, this scenario is quite unlikely given that the tube had lasted for 23 years.

6.4.4 Manner of Failure and Conclusion

The tube had burst during a period of gross overheating, reaching at least 830°C, aided by the presence of deep pre-existing cracks, caused by

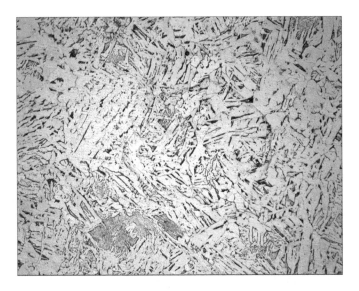

FIGURE 6.21
Specimen X1: Shows that there was a less darkish phase than at the burst location. The dark phase was mainly pearlite but some ferrite/carbide phases were also present. (The microstructure also consisted of coarse Widmanstatten ferrite with some darkish phase.) Mag 200×.

intergranular Cu penetration of an unusual nature. The Cu penetration had likely been due to the presence of Cu-rich deposits, which had diffused into the steel when it became grossly overheated. However, even in the absence of the Cu penetration and cracking, it is very likely that the tube would have eventually burst from the severe overheating. The overheating could only have been caused by some water starvation. There was no obvious evidence of any inadequacy in operation or maintenance. None of the supplied tube samples showed undue thinning or abuse in service as proven by the laboratory examination.

6.5 Case Study 3: Boiler Tube Failure at a Power Plant

Failure analysis was conducted on the ruptured reheater swage tube from a boiler in a power plant. The rupture was caused by creep cavitation, which is a function of tensile stress, temperature, and time. The rupture location had a heavily heat-degraded microstructure and had thinned down due to a combination of metal loss at the internal and external surfaces, and plastic deformation. Internal oxidation was due to steam oxidation and the very thick steam oxides present indicated that temperatures had substantially exceeded values considered safe for T23 material. Thick internal oxides reduced heat conduction to the steam resulting in runaway overheating of the metal. External oxidation was caused by attack by the hot flue gases and ash. The dark and adherent top layer suggested some abnormality in the ash composition that was not present in the early stages of operation [7]. Plastic deformation occurred in the later stages of the creep rupture process when the remaining strength could no longer sustain the applied stress. The material hardness at this location had dropped down to between 116 and 121 HV300g, which is way below acceptable limits.

6.5.1 Background Information

A tube leak was detected around the vertical reheater tube of a boiler unit. The leak was at a rupture at swage Tube no. 8 of Panel 8. Failure analysis was performed on this tube. It was first examined visually; then samples were taken from the ruptured tube at the location of the failure and locations away from the failure for metallographic examination and a hardness test. The boiler design parameters are shown in Table 6.2.

6.5.2 Visual Examination

Figure 6.22 shows photographs of the boiler tube containing the rupture, about 104 cm in length. The tube length was comprised of two sections, called *T91* and *T23* following the names of their materials. Section T91 was welded to T23, and they were roughly 27 and 87 cm in length, respectively. The weld joint was made at the swaged end of the T23 section.

Figure 6.22a shows the front view of the failed location, and Figure 6.22b shows the rear view. The failure occurred at the T23 side, approximately 15 cm from the weld joint. The rupture occurred on the front of the tube, facing the gas flow direction. The ruptured part was slightly bulged. Here, thick layers of adherent and hard external oxides could be seen; the outer

TABLE 6.2

Operating and Design Parameters

No.	Description	Parameter
1	Load	700 MW
2	Temperature at the main steam pipe	542°C
3	Pressure at the main steam pipe	164 bar
4	Metal temperature at the reheater outlet header	Range: 530–650°C
5	Design temperature at the swage tube area	595°C
6	Design pressure at the swage tube area	50 bar
7	Design OD at the swage tube area	(i) 50.8 mm, (ii) 57.0 mm
8	Design thickness at the swage tube area	(i) 3.50 mm, (ii) 3.50 mm
9	Minimum allowable thickness at the swage tube area	(i) 2.69 mm, (ii) 2.40 mm
10	Material at the swage tube area	ASME-SA 213 T23

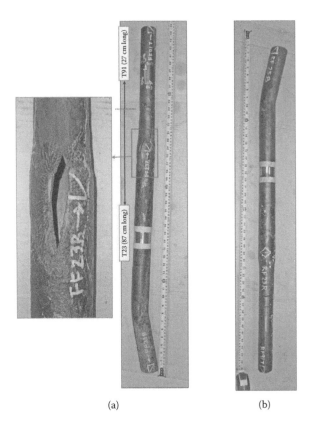

FIGURE 6.22
(See Colour Insert.) (a) The front view of the failed location with a close-up view of the rupture. (b) Rear views of the failed location. The failure had occurred at the T23 side. The oxides around the bulge had cracked. The locations of the metallographic specimens FF23B, FF23R, FF91T, RF23B, RF23R, and RF91T are marked on the tubes.

layers were darkish in appearance but the inner layers were reddish brown. The oxides around the bulge had also cracked.

Figure 6.23a,b are views of the ruptured part seen from the left and right part of Figure 6.22a, respectively. These figures show clearly the thick internal and external oxides at the rupture edge. The metal thickness at the ruptured location was less than 1 mm.

6.5.3 Metallographic Examination and Hardness Test

Axial sections were excised from the front and rear parts of the ruptured plane and also from the front and rear parts of the T23 and T29 tube ends. The positions of the metallographic specimens FF23B, FF23R, FF91T, RF23B, RF23R, and RF91T are marked on the tubes as shown in Figure 6.22. The letters 'FF' and 'RF' denote the front face and rear face, respectively. Arrow 'SP' in

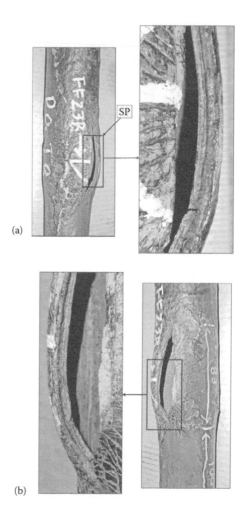

(a)

(b)

FIGURE 6.23
(a) The ruptured part viewed from the left part of Figure 6.22a. The ruptured edge had thick internal and external oxides and the metal thickness was less than 1 mm. Arrow 'SP' points to the section plane that was cut out for metallographic examination. (b) The ruptured part viewed from the right part of Figure 6.22a. The ruptured edge had thick internal and external oxides and the metal thickness was less than 1 mm.

Figure 6.23a points to the section plane that was cut out for metallographic examination. All these sections were subjected to a micro-hardness test and metallographic examination under an optical microscope. Results of the investigation are outlined below.

    a. Specimen FF23R

        This specimen was located at the edge of the rupture. Figure 6.24 (at 75× magnification) shows an axial section along plane SP of

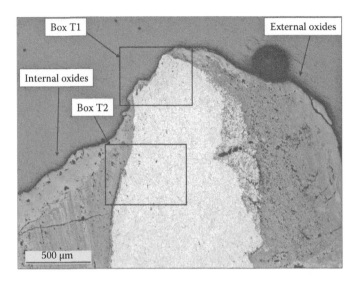

FIGURE 6.24

The axial section along plane SP of Figure 6.23a. The metal thickness at the tip was only about 600 microns; the external oxides were approximately as thick and the internal oxides were even thicker. The areas in Boxes T1 and T2 are magnified in Figures 6.25 and 6.26, respectively. Mag. 75×.

Figure 6.23a. The metal thickness at the tip was only about 600 microns; the external oxides were equally thick and the internal oxides were even thicker. The sections in Boxes T1 and T2, about 1 mm apart, are magnified at 300× and shown in Figures 6.25 and 6.26, respectively.

The microstructure (Figure 6.25) consisted of fairly fine-grained tempered bainite with carbide precipitates and creep cavities. The grains were seen to have stretched in the circumferential direction. The dark patches enclosed in circles are creep cavities. Figure 6.26 (magnified view of Box T2 from Figure 6.24) shows the same features as the ones in Figure 6.25, except that there is less grain stretching here.

The area of Box T3 in Figure 6.26 is further magnified for examination. The creep cavities were seen to have formed mainly at the grain boundaries. Carbide agglomeration and coarsening had proceeded to an advanced degree. The micro-hardness at this location ranged between 116 and 121 HV300g.

b. Specimen RF23R

This specimen comes from the rear of the tube, located 180° from the rupture site and away from the gas flow. The thickness of this part was roughly 3.5 mm. The microstructure in the centre section (Figure 6.27a, magnified 150×) shows fairly fine-grained, tempered bainite with numerous carbides present. The enlarged view (Figure 6.27b) shows that the carbide agglomeration and coarsening

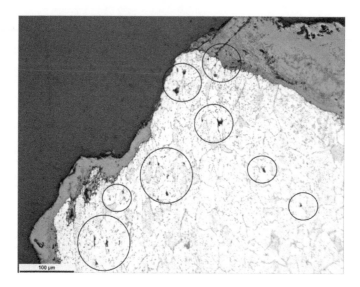

FIGURE 6.25
Magnified view of Box T1 from Figure 6.24. The microstructure comprised fairly fine-grained, tempered bainite with carbide precipitates and creep cavities. The grains appeared to have stretched; the dark patches that are enclosed in the circles are creep cavities. Mag. 300×.

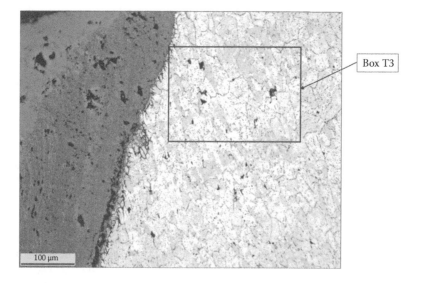

FIGURE 6.26
Magnified view of Box T2 from Figure 6.24. The same features as in Figure 6.25, except that there was less grain stretching here. Mag. 300×.

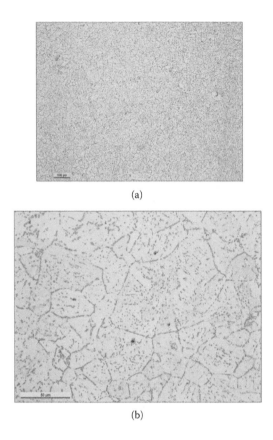

(a)

(b)

FIGURE 6.27
(a) Located at the rear of the tube, away from the gas flow, at the centre of the section: The microstructure comprised fairly fine-grained, tempered bainite with numerous carbides. Mag. 150×. (b) Magnified view of the centre part of (a) shows that carbide agglomeration and coarsening was fairly advanced, but not to the level as shown in Figure 6.26. Mag. 750×.

were fairly advanced, but not to the level shown in Figure 6.26. The micro-hardness around this site was between 151 and 165 HV300g.

c. Specimen FF23B

Specimen FF23B was taken 72 cm from the rupture site at the front face. The microstructure of the external surface (Figure 6.28a, at 150× magnification) was comprised of fairly fine-grained, tempered bainite with numerous carbides. Darkish spots were present but could not be identified at this magnification. Figure 6.28b is an enlarged photograph (magnified 750×) of the centre part of Figure 6.28a. Carbide agglomeration and coarsening were observed to be less advanced than in Figure 6.27b. Dark patches were identified as inclusions or pulled out inclusions, not creep cavities. The micro-hardness at this location ranged between 159 and 166 HV300g.

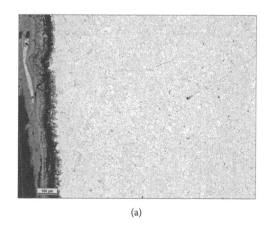

(a)

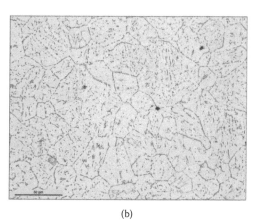

(b)

FIGURE 6.28

(a) T23 tube, about 72 cm below the rupture plane, at the front of the tube facing the gas flow, near the external surface: The microstructure comprised fairly fine-grained, tempered bainite with numerous carbides. Darkish spots were present that could not be resolved at this magnification. Mag. 150×. (b) Magnified view of the centre part (a). Carbide agglomeration and coarsening was less advanced than in Figure 6.27. Dark patches were identified as inclusions or pulled out inclusions, and not creep cavities. Mag. 750×.

   d. Specimen RF23B

   This specimen was located 72 cm from the rupture site at the rear face. The microstructure near the external surface (Figure 6.29a, magnified 150×) also comprised fairly fine-grained, tempered bainite with numerous carbides. Darkish spots were observed with 750× magnification at the centre part of Figure 6.29a. Carbide agglomeration and coarsening were found (Figure 6.29b) to be less advanced than in Figure 6.28b. The dark patches are not creep cavities but inclusions. The micro-hardness at this location ranged between 170 and 178 HV300g.

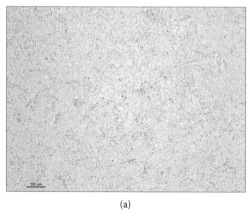

(a)

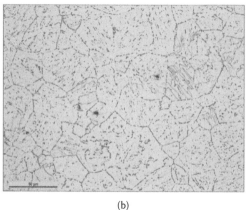

(b)

FIGURE 6.29

(a) T23 tube, about 72 cm below the rupture plane, at the rear of the tube away from the gas flow, near the external surface: The microstructure comprised fairly fine-grained, tempered bainite with numerous carbides. Darkish spots were present that could not be resolved at this magnification. Mag. 150×. (b) Magnified view of the centre part (a). Carbide agglomeration and coarsening was less advanced than in Figure 6.28. Dark patches were identified as inclusions or pulled out inclusions, and not creep cavities. Mag. 750×.

e. Specimens HV91T and RF91T

The examined specimens HV91T and RF91T were located at about 42 cm from the rupture site at the front face and rear surface, respectively. The microstructures near the external surface of both specimens comprised fairly fine-grained, tempered martensite with numerous carbides. Darkish spots were present, and carbide agglomeration and coarsening were more advanced in HV91T than in RF91T. The dark patches were not creep cavities but inclusions. The micro-hardness varied between 178 and 187 HV300g for HV91T, and between 203 and 211 HV300g for RF91T.

6.5.4 Discussion and Conclusions

The investigation revealed that the rupture had occurred by creep cavitation, which is a function of tensile stress, temperature, and time. The rupture location had a heavily heat-degraded microstructure and had thinned down due to a combination of metal loss at the internal and external surfaces, and plastic deformation.

Internal oxidation was due to steam oxidation. The presence of the very thick steam oxides indicates that temperatures had substantially exceeded the safe limit specified for T23 material. Thick internal oxides reduced heat conduction to the steam and resulted in a runaway overheating of the metal. External oxidation was caused by attack by the hot flue gases and ash. The dark and adherent top layer suggests some abnormality in the ash composition that was not present in the early stages of operation. Plastic deformation occurred in the later stages of the creep rupture process when the remaining strength could no longer sustain the applied stress. The hardness of the material at this location fell to between 116 and 121 HV300g, which is well below acceptable limits.

The Tube T23 located 72 cm below the rupture site suffered some heat degradation but there were no creep cavities found. The hardness level at the external surface on the front face was between 159 and 166 HV300g, a borderline between 'accept' and 'reject'.

The T91 tube located 42 cm above the rupture site also suffered from some heat degradation but no creep cavities were visible. The hardness level at the external surface on the front face was between 178 and 187 HV300g, again a borderline between 'accept' and 'reject'.

From the hardness patterns at the ruptured region and at the locations 72 cm below and 42 cm above, it is clear that the overheating had been quite localized, likely due to some flue gas channeling.

6.6 Case Studies 1, 2, and 3: Concluding Remarks

Boiler system failures in conventional boilers can be prevented if sufficient preventive maintenance and efficient operation monitoring are practiced. In the first incident, failure was due to water disruption because water controllers had failed to work in which malfunctioned relays/switches did not turn off the burner. This would not have happened if the control components were serviced regularly. The second incident was also a matter of overheating facilitated by deep pre-existing cracks through intergranular copper penetration. The root cause for overheating is likely to be water starvation. The third case involved a modern boiler which blew up despite the use of high heat-resistance T23 grade material. The rupture was quite localized in nature and had likely been caused by flue gas channeling. This is difficult to detect by instrumentation and the only remedy is to ensure that the the gas flow paths are not blocked and that external deposits are cleaned off regularly.

To conclude, the three case studies highlight the importance of boiler operation, handling, and maintenance. A boiler failure investigator and analyst must have knowledge of all the metallurgical failure modes in addressing a problem. There is ample literature on the subject which the investigator could use and apply. Nevertheless, each boiler and its problem might be unique. Modern boilers today are designed for high performance and capacity, and apply very high-quality materials. They demand expert engineers in their design of special features, as well as experienced and competent system failure analysts.

Problems and Answers

Problem 6.1

A boiler tube made of ASME SA-213 Grade T2 contains 0.15% C and 0.6% Cr. The tube failed in operation with a fishmouth fracture typical of overheating. The tube had been operating normally for three years at a steam temperature of 420°C before failure occurred. A metallographic examination at various locations revealed the following microstructures:

(a) At a distance of 5000 mm from the fracture: Ferritic/pearlitic microstructure with lightly spheroidized pearlite (heat degraded)
(b) At the thinned lip of the fishmouth: 100% martensite/bainite
(c) At a location 200 mm away from the fracture: 50% martensite/bainite with 50% heavily heat-degraded ferrite/pearlite

Explain how the microstructures above had formed. Assume that the eutectoid diagram has the following coordinates:

(i) At C = 0%, A3 = 900°C
(ii) Eutectoid point is at T = 750°C and C = 0.7%
(iii) Maximum solubility of 0.02 wt% carbon in ferrite

Answer 6.1

(a) The microstructure is typical of a ferritic/pearlitic steel (T2) after some years of operation at about 450°C (metal temperature = steam temperature + 30°C); normal.
(b) The metal temperature had exceeded the A3 temperature. In this case, the austenite that formed had transformed to the non-equilibrium

microstructures martensite and bainite through fast cooling. This occurred when the high-temperature, high-pressure steam escaped from the burst tube and expanded adiabatically, resulting in a sudden drop of temperature and pressure.

Note: The phase diagram below is not drawn to scale.

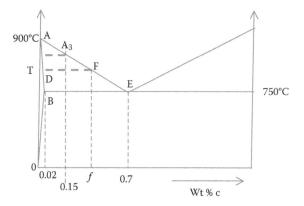

The A_3 temperature can be estimated by constructing the phase diagram based on the information given. For easy estimation, assume AE and AB to be straight lines so that the linear estimation of the temperature can be applied. The A_3 temperature is estimated to be 867.9°C with the following calculation:

$$\frac{900 - A_3}{900 - 750} = \frac{0.15 - 0}{0.7 - 0}$$

$$A_3 = 900 - \frac{0.15}{0.7} \times 150 = 867.9°C$$

(c) At this condition, 50% of the material had transformed to austenite and the remaining 50% was just below the eutectoid temperature. A tie line DF is constructed in the phase diagram. Since composition at B is only 0.02 wt% C, which is very small, and composition at E is even closer to zero wt% C, it can be assumed that the composition at E to be zero wt% C. To estimate the temperature reached (T), first, the composition f is estimated with the lever rule:

$$\frac{f - 0.15}{f - 0} = \frac{0.5}{1}$$

$$f = 0.3 \text{ wt\% c}$$

Then, using linear estimation, the max temperature reached T can be obtained as follows:

$$\frac{900 - T}{900 - 750} = \frac{0.3 - 0}{0.7 - 0}$$

$$T = 900 - \frac{0.3}{0.7} \times 150 = 835.7°C$$

The metal temperature at 200 mm away from the fracture had reached about 836°C.

Problem 6.2

You are an engineer in charge of a power boiler. Your task is to ensure that the steam path is in good condition and passes the periodic inspections by regulatory bodies, for a license to operate the boiler. What inspections/tests will you perform?

Answer 6.2

The following procedures should be used:

- Firstly, perform a visual inspection to ensure that none of the components (tubes and headers) have suffered from visible changes in shape or positioning, and that the flue gas path is clear. Such occurrences call for immediate action.
- Perform a non-destructive examination (NDE) at critical locations. Non-acceptable indications call for immediate action.
- Perform tube dimensional measurements, hardness testing, and a microstructure evaluation on critical locations, for long-term trending.
- Hardness testing should be performed by qualified personnel using approved portable hardness testing machines.
- A microstructure evaluation can be performed in a non-destructive manner with the use of plastic replicas. The replica slides must be examined and evaluated by qualified personnel.
- The results should be trended and compared throughout the life of the components. Excessive thinning rates can be detected and components can be changed or repaired before thickness levels fall

below acceptable values. Excessive thermal degradation that can lead to creep failures can be detected by changes in hardnesses and microstructures and appropriate action can be taken.

Problem 6.3

An investigator conducts a study on the heat transfer and the effects of internal scale thicknesses on a reheater tube metal temperature of a boiler, which operates with a steam temperature 530°C under pressure 13,700 kPa. The boiler tube has an OD of 2.0 inch with a 0.2 inch wall thickness, made of steel grade SA-213 T-22.

Using the following information [10]:

Thermal conductivity of tube metal, k_T: 28.884 W/m.K

Thermal conductivity for scale, k_D: 0.592 W/m.K

Flue gas temperature, T_O: 800°C

Steam side heat transfer coefficient, h_S: 1702 W/m²K

Gas side heat transfer coefficient, h_0:   92 W/m²K

Determine the temperature rise in metal temperature due to the increase in scale thickness from 30 microns to 500 microns. The overall heat transfer coefficient is given by [10]:

$$U_0 = \left(\frac{1}{h_O} + \frac{r_O \ln(r_O/r_i)}{k_T} + \frac{r_o \ln(r_i/r_S)}{k_D} + \frac{r_O}{r_S h_S} \right)^{-1}$$

where the subscript 'O' denotes the gas side, "i" denotes the inner surface of the tube, "S" denotes the steam side, "T" denotes the tube, and "D" denotes the scale deposit. The temperatures and radii are illustrated in the following schematic diagram of a boiler tube section.

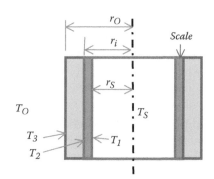

Answer 6.3

$Ts = 530°C$, original tube wall thickness = 0.2 in

$r_o = 2.0$ in$/2 = 1.0$ in = 25.4 mm

$r_i = r_o - 0.2$ in = 0.8 in = 0.8×25.4 = 20.32 mm (before scale is formed)

It is assumed that when the scale is formed through a reaction between the steel and steam, the metal thickness reduces by half the thickness of scale. Therefore, for 30 micron scale thickness,

$$r_i = 20.32 + 0.030/2 = 20.335 \ mm$$

$$r_S = 20.335 - 0.030 = 20.305 \ mm$$

The overall heat transfer coefficient,

$$U_0 = \left(\frac{1}{h_O} + \frac{r_o \ln(r_0/r_i)}{k_T} + \frac{r_o \ln(r_i/r_S)}{k_D} + \frac{r_O}{r_S h_S} \right)^{-1}$$

$$= \left(\frac{1}{92} + \frac{25.4 \times 10^{-3} \ln\left(\dfrac{25.4}{20.335}\right)}{28.884} + \frac{25.4 \times 10^{-3} \ln\left(\dfrac{20.335}{20.305}\right)}{0.592} + \frac{25.4}{20.305 \times 1700} \right)^{-1}$$

$$= 84.292 \ W/m2K$$

The heat flux through the tube wall,

$$q'' = \frac{Q}{A_O} = U_0(T_0 - T_S) = 84.292(800 - 530) = 22.759 \ kW/m^2$$

The metal temperature at the outer surface and inner surface are T_3 and T_2, respectively.

The heat flux $q'' = \dfrac{Q}{A_O} = \left(\dfrac{T_0 - T_3}{1/h_O} \right)$

Therefore, $T_3 = T_0 - q''/h_O = 800 - 22759/92 = 552.6°C$

The heat flux $q'' = \dfrac{Q}{A_O} = \left(\dfrac{T_3 - T_2}{\dfrac{r_o \ln(r_0/r_i)}{k_T}} \right)$

$$T_2 = T_3 - q'' \frac{r_o \ln(r_o/r_i)}{k_T} = 552.6 - (22759) \frac{25.4 \times 10^{-3} \ln\left(\dfrac{25.4}{20.335}\right)}{28.884}$$

$$= 552.6 - 4.5 = 548.1°C$$

For 500 micron scale thickness, the metal thickness reduces by half the thickness of scale 250 microns from the inner surface of the tube.
Therefore,

$$r_i = 20.32 + 0.25 = 20.57 \; mm$$

$$r_S = 20.57 - 0.5 = 20.07 \; mm$$

The overall heat transfer coefficient,

$$U_0 = \left(\frac{1}{h_o} + \frac{r_o \ln(r_o/r_i)}{k_T} + \frac{r_o \ln(r_i/r_S)}{k_D} + \frac{r_o}{r_S h_S} \right)^{-1}$$

$$= \left(\frac{1}{92} + \frac{25.4 \times 10^{-3} \ln\left(\dfrac{25.4}{20.57}\right)}{28.884} + \frac{25.4 \times 10^{-3} \ln\left(\dfrac{20.57}{20.07}\right)}{0.592} + \frac{25.4}{20.07 \times 1702} \right)^{-1}$$

$$= 77.794 \; W/m^2K$$

The heat flux through the tube wall,

$$q'' = \frac{Q}{A_O} = U_0 (T_0 - T_S) = 77.794(800 - 530) = 21.004 \; kW/m^2$$

The heat flux $q'' = \dfrac{Q}{A_O} = \left(\dfrac{T_0 - T_3}{1/h_O} \right)$

$$T_3 = T_0 - q''/h_O = 800 - 21004/92 = 571.7°C$$

The heat flux $q'' = \dfrac{Q}{A_O} = \left(\dfrac{T_3 - T_2}{\dfrac{r_o \ln(r_o/r_i)}{k_T}} \right)$

$$T_2 = T_3 - q'' \frac{r_o \ln(r_o/r_i)}{k_T} = 571.7 - (21004) \frac{25.4 \times 10^{-3} \ln\left(\dfrac{25.4}{20.57}\right)}{28.884}$$

$$= 571.7 - 3.9 = 567.8°C$$

Summary of temperature and the temperature rise:

Scale Thickness	Temperature T2 (tube inner surface)	Temperature T3 (tube outer surface)
30 microns	548.1°C	552.6°C
150 microns	567.8°C	571.7°C
Temperature rise	**19.7°C**	**19.1°C**

Note to reader: This problem is just to show that excessive steam-side oxides cause metal temperatures to rise and hasten creep failure if not removed. In fact, power plants do check on oxide thicknesses regularly and carry out chemical cleaning when there is excessive thickness.

Problem 6.4

A metallurgical examination on the microstructure of boiler tubes helps to understand the heating process that the materials have undergone. Explain the possible heating processes the materials could have undergone if the following microstructure is identified in a failure analysis.

(a) Widmanstatten ferrite
(b) Degraded pearlitic-ferritic microstructure

Answer 6.4

(a) Widmanstatten ferrite: Widmanstatten ferrite is formed when austenite in steel cools down from above A3 temperatures, at rates between that required for normal ferrite/pearlite formation and bainite/martensite formation. The localized presence of Widmanstatten ferrite at tube ruptures thus proves overheating to A3 temperatures. See Bodnar and Hansen [8], Effects of Austenite Grain Size and Cooling Rate on Widmanstatten Ferrite Formation in Low-Alloy Steels, for more information about Widmanstatten ferrite.

(b) Degraded pearlitic-ferritic microstructure: Carbide precipitation or cementite spherodization, followed by carbide coarsening and agglomeration occurs by diffusion at a high temperature (above 500°C); the materials soften and strength is reduced; the grains can be observed to be stretched if under a certain stress level. This microstructure indicates that the materials had been exposed to a high temperature (above 500°C), and the temperature could be estimated depending on the severity of the thermal degradation, the grade of steel, and the period of exposure. See Cvetkovski et al. [9] Thermal Degradation Pearlitic Steels: Influence on Mechanical Properties Including Fatigue Behaviour, for more information about thermal degradation of pearlitic steel.

Problem 6.5

(a) The two steels, ASME SA-213 T9 and T91, appear to have almost the same chemical compositions, but the creep performance of T91 is superior. Explore this statement further and explain why T91 is so superior.

(b) Boiler tubes made of austenitic stainless steels (ASSs) are known to be able to withstand higher temperature service than steels with ferritic, pearlitic, bainitic, or martensitic microstructures. What are the reasons for this?

(c) From your answers to (a) and (b) above, what do you think would be the basic principles in the metallurgical design of alloys with creep strengths higher than the basic ASS?

Answer 6.5

(a) The basic chemical compositions of the steels, in %, are:

T9: C 0.15 max; Mn 0.30–0.60; Si 0.25–1.00; Cr 8.00–10.00; Mo 0.90–1.10

T91: C 0.07–0.14; Mn 0.30–0.60; Si 0.20–0.50; Ni 0.40; Cr 8.00–9.50; Mo 8.0–9.5; V 0.18–0.25; Nb 0.06–0.10; N 0.030–0.070; Al 0.02; others are Ti 0.01, Zr 0.01

The maximum allowable design stresses specified by ASME for creep rupture lives of 100,000 hours at 600°C are 20.7 and 65.0 MPa, respectively, for T9 and T91. T91 is demonstrably superior and would translate to a thickness and weight advantage of more than 300% over the T9 for the same application.

The composition of T9 allows it to have a fairly coarse ferritic/pearlitic microstructure if annealed or a martensitic microstructure if normalized (the martensite would of course have to be tempered before use). For creep-resisting purposes, the martensite would not be superior. (Note to readers: Compare the performance of T9 with T22.)

The major alloying elements of T91 (Cr and Mo) do not differ much from T9, but T91 has some very important micro-alloying elements such as V, Nb, and N, which form fine precipitates such as carbides and carbonitrides that are stable at high temperatures. T91 is used after normalizing and tempering, in which condition, it has a fine martensite microstructure with a very high-dislocation density and large quantities of fine precipitates. The precipitates lock up the dislocations and grain boundaries, hence reducing the speed of thermally induced diffusion processes responsible for creep. This effect persists at conditions (time, temperature, and stress dependent) much more severe than what the T9 can withstand.

(b) There are two aspects to high-temperature performance of ASS, namely, oxidation resistance and creep resistance. With respect to the first, ASSs have higher Cr content than the non-ASS and it is this Cr that confers oxidation resistance. In the second respect, when the temperature of the non-ASS exceeds the A1 temperature, transformation to the austenite phase will ensue and the creep strength will quickly drop, usually resulting in a rupture. This loss in creep strength occurs because the austenite of the non-ASS has considerably lower creep strengths than the other microstructures. On the other hand, the ASS already started with an austenitic microstructure and thus no distinct phase change occurs up to the melting point (though undesirable phases may precipitate after certain temperatures). However, the austenite of ASS has a higher creep strength than non-ASS steels at temperatures beyond A1 because it contains much higher quantities of alloying elements.

(c) A strong, stable austenitic matrix with large quantities of temperature-stable precipitates would be required. The matrix would need to contain large proportions of austenite formers such as Co and Ni and the precipitates could be various forms of intermetallics and carbides or carbonitrides. Such development would lead us to the superalloys.

References

1. Callister, W.D., *Materials Science and Engineering, An Introduction*, 3rd ed., New York: John Wiley & Sons, 1994.
2. Jones, D.R.H., Creep Failures of Overheated Boiler, Superheater and Reformer Tubes, *Engineering Failure Analysis*, Vol. 11, No. 6, December 2004, 873–893.
3. Meadowcroft, D.B., High-Temperature Corrosion and Coating in Oil- and Coal-Fired Boilers, *Materials Science and Engineering*, Vol. 88, April 1987, 313–320.
4. EPRI Report TR-102433-V2, *Boiler Tube Failure Metallurgical Guide*, Vol. 2: Appendices, Palo Alto, CA: EPRI, 1993.
5. Michél, J., M. Buršák, and M. Vojtko, Microstructure and Mechanical Properties Degradation of CrMo Creep, *Mechanical Engineering Materiálové Inžinierstvo*, Vol. 18, 2011, 57–62.
6. Spurr, J.C., Significance of Copper Deposits Associated with a Boiler Tube Failure, *Anti-Corrosion Methods and Materials*, Vol. 6, No. 8, 1959, 233–237; Hargrave, R.E., Unusual Failures Involving Copper Deposition in Boiler Tubing, *Corrosion*, Vol. 47, No. 7, 1991, 555–567.
7. Srivastava, S.C., K.M. Godiwalla, and M.K. Banerjee, Review Fuel Ash Corrosion of Boiler and Superheater Tubes, *Journal of Materials Science*, Vol. 32, 1997, 835–849.

8. Bodnar, R.L. and S.S. Hansen, Effects of Austenite Grain Size and Cooling Rate on Widmanstatten Ferrite Formation in Low-Alloy Steels, *Metallurgical and Materials Transactions A*, Vol. 25A, 1994, 665–675.

9. Cvetkovski, K., J. Ahlstrom, and B. Karlsson, Thermal Degradation Pearlitic Steels: Influence on Mechanical Properties Including Fatigue Behaviour, *Materials Sc. Tech.*, Vol. 27, 2011, 648–654.

10. French, D.N., *Metallurgical Failures in Fossil Fired Boilers*, 2nd ed., New York: John Wiley & Sons, 1992.

7

Infrastructure Failure Analysis

7.1 Introduction

The construction of large-scale infrastructures involves huge capital investment, and if it does not follow rigorous design and construction procedures, various forms of failure may arise. In the worst-case scenario, the structure may even collapse. Common modes of structural failure include cracks, deformation, corrosion, ground settlement, fatigue, and fracture. Improper selection of materials, overloading, faulty design, and natural calamities are some of the contributing factors.

This chapter presents two case studies in failure analysis of huge buildings. The first case concerns the problem of corrosion under insulation (CUI) of a metal-based roof of a large building. The problem first started when water entered the insulation section of the roof during its construction. However, the architectural design of the roof only permitted very little indirect air ventilation. As a result, water trapped in the glass wool and cement boards could not dry in time. Subsequently, the roof structure became corroded.

The second case study involves condition assessment and monitoring work conducted in a college. This work began when some cracks appeared on the brick walls and structural foundation several years after the building was completed. Since then, two tubewells were constructed 100 meters from the Administration and Academic faculty so that underground water could be obtained for construction purposes. And further, the cracks in the block of buildings became larger and new cracks appeared, the door and window frames became distorted, the glass window panels fractured, and the floor slabs dipped.

7.2 Case Study 1: Corrosion under Insulation of a Metal-Based Roof

7.2.1 Background Information

In a metal-based roof of a very large building in Malaysia, water was discovered to have entered the roof during its construction. The roof covered an

(a)

(b)

FIGURE 7.1
Two photographs of the roof construction.

area of 57,500 m^2 and it was made of multi-layers of insulating cement board and glass wool. The design was such that the roof was partially enclosed, permitting very little free airflow or ventilation. Thus, the water trapped inside could not dry off quickly causing severe corrosion under insulation (CUI) at one of the roof load-bearing components.

Figure 7.1a shows a quarter section of one bay, with the initial steel support framework laid; the 1st layer cement board is fixed to this framework. Figure 7.1b shows a similar section with the Z-spacers (the horizontal strips in the figure) installed on top of the 2nd layer cement board. The 3rd layer cement board will rest on the spacers. Figure 7.2 shows a cross-section of the roof. The following is a description of how the roof was constructed.

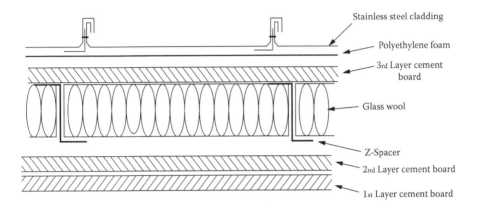

FIGURE 7.2
Schematic diagram of the cross-sectional view of the roof.

- C-purlins are attached to the supporting framework in a parallel fashion and at an angle of 45 degrees to the direction of the central pan. Single and double purlins are installed at a distance of 610 mm apart in other sections of the roof.

- Two layers of cement boards (the 1st and 2nd layer cement boards) are then bolted onto the top of the purlins using self-tapping screw nails.

- Galvanized Z-spacers are placed on top of the 2nd layer cement board at a distance of 610 mm apart and at an angle of 45 degrees to the C-purlins. A Z-spacer is a Z-shape 1.6 mm-thick steel strip coated with a layer of 20 μm zinc on both sides. It is used in the roof to separate as well as install the insulation materials. Because the Z-spacer is coated with zinc, it is called *galvanized steel Z-spacer*. Self-tapping screws are driven into the points of intersection with Z-spacers joining the two layers of cement boards with the purlins. The screws measure 6 × 48 mm. A total of 1406 screws are used per bay. Glass wool is then inserted between the Z-spacers.

- The 3rd layer cement board is screwed onto the top of the Z-spacers. A polyethylene foam sheet covers the top of the 3rd layer cement board.

- Stainless steel L-clips are then placed on top of the polyethylene sheet, perpendicular to the Z-spacers at a distance of 440 mm apart. A 3rd layer cement board is nailed onto the L-clip and Z-spacer intersection. The screw nails measure 5 × 40 mm and a total of 2738 pieces are used per bay.

- Next, long sheets of the stainless steel cladding are placed between the L-clips. The claddings are bent 90 degrees to fit the vertical edge of the clips. The edges of the adjacent sheets are welded to the enclosed clip to make a watertight joint. The protruding edges of the sheets beyond the weld joint are folded into an inverted U.

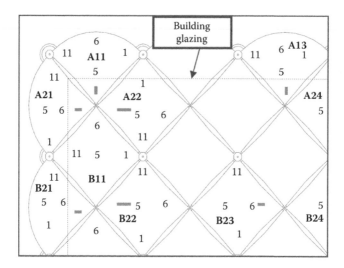

FIGURE 7.3
Layout of the affected bays and the positions of the monitoring probes (to be discussed in Section 7.2.6). Numbers 1, 5, 6, and 11 denote positions of probes 1, 5, 6, and 11, respectively. Ambient probes are located at bays B1-1, D12-2, and D11-4 (the last two are not shown in the figure).

The Z-spacer on the roof has many functions:

- Provides space for the insulating materials
- Provides support points for installation of the roof cladding
- Supports downward loading for the C-purlin and upward wind suction for the cladding

Figure 7.3 displays the roof bay configuration. The roof is composed of 93 segments. Each segment is called a *bay*. The rows are designated as zones A1, A2, B1, B2, ... E3, and the columns are numbered 1, 2, ... 6.

7.2.2 Identification of the Problem

Apparently, water had entered the insulation zone of the roof during its construction and became trapped in the glass wool between the 2nd and 3rd layer cement boards, wetting the Z-spacers. The wool was drenched in many locations and could not dry out in the enclosed space. In addition, fresh water entered through small leaks in the steel cladding. This all resulted in fast corrosion of the Z-spacers, which is known as corrosion under insulation (CUI). If prolonged, the corrosion will lead to fast disintegration of the Z-spacers.

CUI is characterized by localized corrosion and pit formation. The extent of the CUI in this case was severe. It was obvious that the water in the roof had to be removed to low enough levels so as not to cause continued corrosion.

Barnhart [1] cautions that water in the insulation could cause deterioration and eventual structural damage by rot, corrosion, or water expansion in freezing water. Ashbaugh and Laundrie [2] have studied the effects that different types of insulation have on the corrosion rate of carbon steel. They found that corrosion under the edge of the insulation was more severe than in the middle part due to free access of water and oxygen. Dharma et al. [3] state that the rate of CUI is determined by the contaminants in the water, heat transfer properties of the metal surface as well as wet/dry conditions of the surface. Other CUI case studies include stainless steel pipes [4,5] and galvanized steel tanks [6].

The building is situated 16.3 m above sea level and about 20 km from the Straits of Malacca. It is a typical coastal location in the tropics. Studies of corrosion rates in such climates appear in many literary works [7–10]. Potgieter-Vermaak et al. [11] have reported many instances of degradation in galvanized iron roofing materials because of tropical weather and increased air pollution. Rates of corrosion depend on relative humidity (*RH*), temperature, moisture condensation, and timing.

In the presence of water, the CUI attack on the Z-spacers was random and localized, the result was the formation of shallow pits and their joining up. Figure 7.4a is a photograph of the corroded Z-spacers taken from the rectified bays in the roof. Figure 7.4b shows the appearance of the surface attack after removing the loose deposits. Figure 7.5 shows a typical cross-sectional profile of some of the pits. The pit depths varied between 0.23 and 1.10 mm in the worst affected bays, compared to the original metal thickness of about 1.6 mm. The attack was considered to be severe, as the roof had only been built a few months earlier. Basically, it did not conform to the contractual obligations.

Obviously, the owner would not accept such a roof and the roofing contractor had to make acceptable rectifications. Due to time constraints, it was only

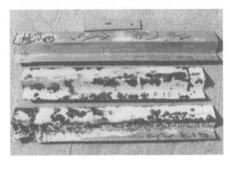
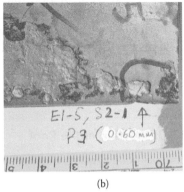

(a) (b)

FIGURE 7.4
(See Colour Insert.) (a) Corroded Z-spacers from rectified bay E1-5 and (b) localized corrosion and pitting attack on the corroded Z-spacer from bay E1-5.

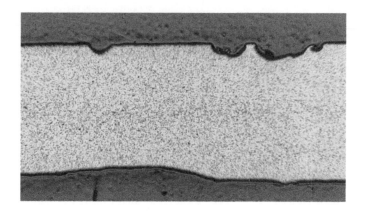

FIGURE 7.5
The corroded Z-spacer (original thickness of 1.6 mm) showing a fairly uniform localized attack on the lower surface of the figure, and an incipient pitting attack on the upper surface.

possible to fully rectify a certain number of the affected bays, the worst ones. The remainder then had to be proven to be fit for service for the designed life of the roof. The steps taken would firstly include the appointment of a third-party expert (TPE), who would oversee the implementation of the proposed program. The program essentially consisted of the following:

(i) An initial thorough inspection of the roof bays to categorize their conditions.

(ii) Verification that the structural performance of even the worst remaining roof will conform to the application standard (BS 5950: Part 1); this involved actual load testing of a model test bay.

(iii) Rectification, comprising: Firstly, of an accelerated drying out operation and confirmation that the bays remained dry; this required long-term monitoring of the temperature and humidity in the space between the 2nd and 3rd layers of the cement boards. Secondly, a research program to identify the corrosion mechanics in such a roof design was needed.

Details of this program are provided in the following sections.

7.2.3 Inspection of the Roof

An inspection program was carried out on the Z-spacers in all 40 bays at the beginning of this study. Nearly all the inspected bays were found to be covered with variable amounts of white and brown deposits, the amount depended upon how wet the bay was. White deposits were zinc oxides from the zinc coating and brown deposits were corrosion products (rust) from the steel substrate.

7.2.4 Load Testing of a Model Test Bay

A model test bay was constructed with Z-spacers in their corroded condition as well as containing various degrees of artificial metal loss, simulated by drilling holes on the flanges and the webs of the spacers. A roof of 8.2 × 8.2 m in size was constructed for the load tests. The structure of the test roof was identical to the roof under study. The model roof was divided into five test bays of 4 m² each, and the spacers were loaded in accordance with BS 5950: Part 1 [12], which allows for a deflection of 2.4 mm over a length of 862 mm. Strains in the spacers were measured by means of electrical strain gauges at critical locations, and converted to stress values. Essentially, the only stresses of high magnitudes were the shear stresses, and these were compared with the yield strength of the spacer material, at 103 MPa. The physical parameters of the Z-spacers at the centre of each test bay were as follows:

i. Test bay A is normal (control bay).
ii. Test bay B is badly corroded (constructed using the corroded Z-spacers removed from rectified bay E1-5, worst-case corrosion).
iii. Test bay C with drilled holes to simulate a 25% loss in area and advanced corrosion.
iv. Test bay D, first with drilled holes to simulate a 35% loss in area and extreme corrosion; and second, an addition of slots cut into the material to simulate the presence of cracks.

For first stage loading, design loads were employed, first without any factors, and then with a factor of 1.4 for wind loading and 1.6 for imposed loading. According to BS 5950, Part I, when checking the strength of a structure, the specified loads should be multiplied by the relevant partial factors and a factor of 1.4 is used for wind loading and a factor of 1.6 is used for imposed loading [12]. For second stage loading, bay D was tested using 250% of the design load and 500% of the suction load. The design loads comprise (i) a downward distributed wind load of 71 kgf/m², (ii) a downward imposed point load of 90 kgf/m², and (iii) an upward distributed wind suction load of 103 kgf/m². Figure 7.6 shows the arrangement for the suction loads.

Table 7.1 is a summary of the test results. The values for the three highest shear stresses and deflections for each bay under each stage of testing are provided.

Results revealed that the severely corroded Z-spacer in test bay B only experienced a maximum of 25.4% of the shear yield strength of 103 MPa, and further, the maximum deflection of –1.62 mm at the factored loads was much smaller than the maximum allowable deflection of 2.4 mm at non-factored loads. These values are very close to those of test bay A, which implies that a Z-spacer with the worst corrosion condition is still very safe for use from

FIGURE 7.6
Loading arrangement for the suction loads.

TABLE 7.1

Simulation Load Test Results

Test Bay/Loading	Maximum Shear Stress (MPa)	Deflection (mm)
A (normal): Stage 1	21.2, 21.8, 22.1	−1.08, −1.09, −1.3
B (corroded): Stage 1	24.3, 24.4, 26.2	−1.58, −1.58, −1.62
C (25% of holes): Stage 1	20.3, 21.6, 21.0	−1.18, −1.23, −1.20
D (35% of holes): Stage 1	53.0, 54.2, 67.9	−2.11, −2.13, −2.23
D (35% of holes with slots): Stage 1	51.3, 51.9, 67.0	−2.02, −2.24, −2.92
D (35% of holes with slots): Stage 2	88.7, 91.7, 98.2	−3.72, −3.92, −4.52

the structural performance point of view. For test bay D with slots, Z-spacers with 35% of holes and Stage 2 loading, was only stressed to 98.2 MPa, less than the shear capacity of 103 MPa, and 250% of the design loads and 500% of the suction load. At these load points, a deflection of −4.52 mm would not cause any breakage in the cement board, or any visible shape deformation in the Z-spacer. This suggests that any further corrosion as serious as a 35% mass loss, if any, would not affect its safety function.

7.2.5 Rectification of the Problems

7.2.5.1 Principles

It has been proven that the Z-spacers in the current condition had a very large margin of safety, but continued CUI may lead to dangerous conditions. The CUI therefore had to be stopped by drying out the roofs and ensuring that there was no further ingress of rainwater; this would then bring the atmospheric conditions back to the normal design conditions; anything less would not be acceptable to the owner. The drying process was carried out by blowing large quantities of hot, dehumidified air into the wet space, from openings made at the top of steel cladding. The dryness condition was monitored by temperature and relative humidity (*RH*) probes inserted through the steel claddings into the blown space. Monitoring was carried out 24 hours per day and the readings were converted to water content and trended. At the same time, test coupons were inserted into the wet space to test for corrosion tendencies. Further, separate small-scale test bays were constructed to simulate actual roof conditions and to obtain data for formulation of corrosion rate laws, which would allow for the prediction of long-term corrosion. The roof would be deemed to be dry when the trend for a specified period of time did not show any increase in water content despite incidents of heavy rain, and also when there was no undue increase in corrosion in the test coupons. The long-term corrosion prediction would provide the confidence that there would be no unexpected corrosion attack.

7.2.5.2 *Methodology*

The investigation for the corrosion problem was carried out via on-site field tests. In addition, 10 test panels of 1.0 × 1.22 m in area were also constructed and installed on the rooftop of a building that would serve as a simulation site. The investigation involved:

i. Corrosion field tests

Metal coupons were inserted into the roof space between the 2nd and 3rd layers of the cement boards on both the roof site and the test panels to assess their weight loss after corrosion.

ii. *RH* and temperature monitoring

The task was to measure relative humidity (*RH*) and temperature (T) and to calculate the water content in both the affected bays and the test panels.

iii. Dry condition criteria

The environmental conditions for which the corrosion rate on the Z-spacer was acceptable could be established based on the results of the test panels obtained in stage (ii).

iv. Formulation of the mathematical models

Two mathematical models were formulated to describe the relationship between the corrosion rate, relative humidity, and temperature based on the results obtained from stage (i) and stage (ii).

v. Analysis and prediction of service life

The models obtained in stage (iv) were used to predict the degree of Z-spacer corrosion during a span of 50 years.

7.2.6 Analysis

The corrosion field tests involved direct measurement of the corrosion rate *in situ* (on-site). The corrosion rate at the different bays on the roof was obtained using the weight loss method. Multiple sets of metal coupons were exposed in the monitoring holes as well as in the roof sections. These coupons were subjected to periodic removal so that the corrosion behavior could be monitored over the exposure time period. The law governing the behavior of corrosion of galvanized steel and steel is given by

$$\Delta W = Kt^{N} \tag{7.1}$$

where W is the weight loss due to corrosion penetration in mm, t is the time of exposure in years, and K and N are coefficient characteristics of the corrosion behavior. K and N were calculated from the corrosion attack figures. Prediction for the long-term performance of the Z-spacers was done based on the correlation figures found in the short-term corrosion data. Five types of metal coupons were used in this study; they were a galvanized steel Z-spacer, a bare steel Z-spacer, a corroded galvanized steel Z-spacer, a flat sheet galvanized steel, and a flat sheet bare steel. It is known in roof environments that galvanized steel, with and without white rust, has a much better corrosion resistance than bare steel. Hence, if the results for bare steel are satisfactory, there is no need to consider those for galvanized steel. Furthermore, there was very little zinc left on the previously corroded Z-spacer as the bare steel substrate was almost completely decayed. As such, the galvanized steel coupons were only intended for control purposes.

Some specimens were taken from the corroded galvanized steel Z-spacer in bays B1–4, C1–4, and E1–5 at the time when they were being repaired. Some samples of galvanized Z-spacer coupons were also obtained from the new Z-spacer strips that were identical to the ones used in the roof construction. The spacer was cold-formed to the shape of a 'Z' using a galvanized hot-roll steel sheet following the Japanese Industrial Standard specification for hot-dip zinc-coated steel sheets and coils (JIS G 3302-1994 [13] grade SGHC Z27). The spacer has an average non-alloyed zinc coating mass of 275 g/m² on both surfaces (equivalent to 27 m of coating thickness on a single surface), minimized spangle, and ordinary chromate treated. They are not smeared with oil or lubricated. The bare steel Z-spacer specimen taken from the base metal in the galvanized Z-spacer was also cold-formed to Z.

All the replicate coupons were produced in accordance to the ASTM Standard Practice for preparing, cleaning, and evaluating corrosion test specimens G1-90 [14]. Before exposure, these coupons were encoded with a stamp mark for future identification. First, they were coarsely sandpapered using a 300-grid sandpaper and then with a finer 600-grid sandpaper. Subsequently, they were scrubbed and cleaned in distilled water, rinsed in methanol, and dried using a blow dryer. Finally, they were carefully weighed out at ±0.01 mg per coupon.

The test coupons were also placed in the monitoring holes and the roof sections. A monitoring hole is a small circular opening (about 70 mm diameter) on the roof. The spaces within the hole can hold up to four types of coupons comprising one unit of one type each. A section is a rectangular opening made on the roof for inserting the larger coupons. Both single and double sections were required. A set of five coupons went into a *single section* of the bay at any one time. Another set of four coupons of five units each, totaling 20, went into the *double section*. All the test coupons were then subjected to a one-time exposure. At the end of the exposure period, each lasting six months, they were completely removed and a new set of panels was put in. The duration of the consultation and commissioned monitoring program was two years. An illustration of a coupon display in a given monitoring hole is shown in Table 7.2, and the bay configuration is shown in Figure 7.3. The set of four coupons comprised a galvanized Z-spacer, a bare steel Z-spacer, a flat sheet galvanized steel, and a flat sheet bare steel.

Separately, the corroded coupons were also placed in the sections measuring 200 × 2000 mm. A11, A13, A21, A25, A26, B21, B23, C23, C25, D112, D115, D122, and D123 were the selected bays for the single sections. Double sections were made at A22, B22, and D126. Figure 7.7 indicates the two types of sections and their coupon displays. The monitoring sections are shown in the photographs in Figure 7.8.

Corrosion was measured using the weight loss method as per ASTM G1-90 standard [14]. This method entailed weighing the coupons first before exposure, and then removing the corroded products after exposure and weighing them again. The difference in weight is a measure of the amount of corrosion developed during the exposure period.

TABLE 7.2

Monitoring Hole Coupon Displays

Bay	Holes	Bay	Holes
A13	1, 5, 6, X2(H11), X3(H11)	C24	X1(H11), X3(H1)
A21	5, 6	C25	1, 5, 11
A22	5, 6, X1(H11), X3(H1)	D112	5, 6, X3(H11)
A26	5, 6, 11	D113	X1(H1), X3(H11)
B11	X1(H1), X3(H11)	D115	6
B14	5	D122	5, 6, X3(H1), X4(H11)
B21	1, 5, 11	D123	5, X1(H1), X3(H11)
B22	5, 6, X1(H11), X3(H1)	D124	X1(H1), X3(H11)
B23	1, 5	D125	X1(H1), X3(H11)
B24	1	D126	5, 6, X1(H1), X3(H11)
C11	X1(H1), X3(H11)	D213	1
C23	X1(H1), X3		

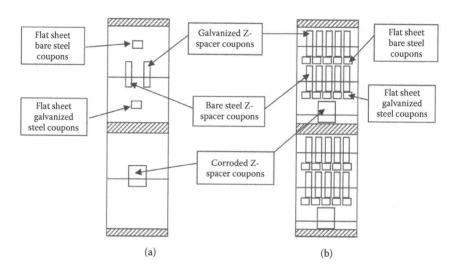

FIGURE 7.7
Coupon orientation in a (a) single section, and a (b) double section.

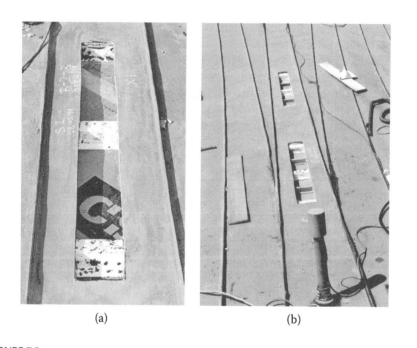

FIGURE 7.8
(See Colour Insert.) (a) A view of the section made on bay B2-3. The Z-spacers were white-rusted and brown rust was observed to have progressed to a significant level. (b) A rectangular section made on bay D12-6. The spacer was corroded and brown rust was obvious.

All the bare steel coupons were cleaned in hexamethylene tetramine solution and hydrochloric acid, as proposed in the ASTM standard procedure. Past experience of corrosion observed in a small sample of exposed coupons showed an ASTM standard ratio of 1:2 mixtures to be appropriate for obtaining a good cleaning curve [13] that indicates a weight loss property. Cleaning was done in an ultrasonic cleaner for about 30 s to 60 s for each cycle. About 5 to 6 cycles were sufficient for the effective removal of the corrosion products in the coupons. The corrosion products in the galvanized coupons were removed using chromium trioxide solution for an immersion time of 60 s at 800°C. Again, the ratio should be 1:2 ASTM standard range.

For the experimental roof test panels, the corrosion rate after the experiment was also determined via the weight loss method. The 10 panels were labeled Panel 1 ... Panel 10. All the test coupons in the roof panels were exposed in acceptably dry condition, except for Panel 7 and Panel 9. The latter were maintained at a high *RH* of 80% and above.

RH and Temperature Monitoring

Temperature and relative humidity are independent parameters that indicate the moisture content in the air. A temperature difference may exist between a surface and its ambient environment. If *RH* in the ambient environment is sufficiently high and the temperature difference causes a surface saturation, then condensation will set in. Therefore, it is important to monitor the relative humidity and the temperature of the test panels so as to determine its water content of the test panels and correlate to the affected bays.

The 30 bays situated at the various sites had to be monitored daily for temperature and relative humidity for a period of two years. Four sensor probes were used to measure *RH* and T at each bay and three probes were used to monitor the ambient air conditions, giving a total of 123 probes. Figure 7.9 is a schematic diagram of a probe mounting.

The probes were inserted into the spaces within the roof through holes made on the roof steel cladding. Each probe contained one temperature sensor and one relative humidity sensor. Each probe was supported by a plastic plug that is mechanically fixed to a hole. *RH* and T readings on the roof bays and the ambient air above the roof were taken every 3 seconds. These were then averaged into half-hour results.

In addition, the monitoring system was also installed on the test panels (Figure 7.10). A single probe was installed at each panel to measure the *RH* and T while two probes were used to measure the *RH* and T of the ambient temperature where the panels were located. A PVC pipe was installed at the right-hand side of the panel which holds a probe that was monitoring the ambient air. Hoses attached to both sides of the panels facilitated free air movement inside these panels.

7.2.7 Discussion of the Results

Experimental test panels were installed in the roof to establish a model for dry conditions at the roof in which the corrosion was low. Figure 7.11 is a

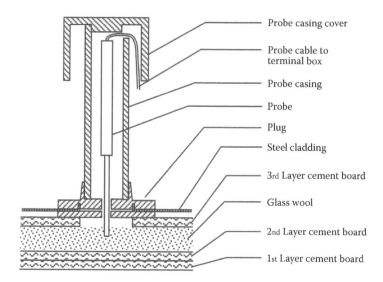

FIGURE 7.9
A schematic diagram of the mounting probe casing onto the roof using a plug.

FIGURE 7.10
Setup of the roof test panels.

record of typical daily results obtained for ambient air, one wet panel and one dry panel on a hot day and on a cold rainy day. Results were taken for relative humidity (*RH*, %), temperature (*T*, °C), and water content (*w*, grams of water content in 1 kg of dry air). The findings (Figure 7.11a,b,c) revealed that the temperature and water content in both the test panels were much higher than that of the ambient air in the daytime. The additional quantity of *w* comes from the release of free water vapor and from water vapor trapped in the cement boards. The *RH* of the wet panel was high, whereas the *RH* of the dry bay was lower than the ambient values in the evening. On a typical cold and rainy day (Figure 7.11d,e,f), *T* and *w* for both panels did not rise very much although *w* in the wet bay was higher than in the dry bay. The *RH* in the wet bay remained very high indeed.

The behavior recorded in the test panels was assumed to be similar to that encountered in the roof space of the large building in this study. From the results of the dry test panels, relative humidity, temperature, and water content for an acceptable or level of roof bay dry condition was then identified. Finding the criteria for an optimal level of *RH*, *T*, and *w* are essential because these values could be used as a reference for evaluating the conditions found in the roof bays watch.

Figure 7.11 reveals that the maximum daily water content (*w*) in the daytime was clearly more prominent. This suggests that in the roof, the cement boards and panels were releasing more water vapour. It provides qualitative information of the amount of water trapped in the cement boards. In addition, corrosion is an electrochemical process that requires the presence of electrolyte. Moisture condensation provided such a medium for corrosion to occur. Condensation was not possible during the daytime because of high temperatures and relatively low humidity. A possible detection of condensation had to be at night or early morning. Results showed that the *RH* was low during the day but it would increase significantly in the evening and then stabilize at pre-dawn around 6:00 A.M. The same 25°C temperature level was also experienced in all the test panels. This is a good basis for comparison of *RH* in both these locations. The variables that best described the conditions occurring in both panels and bays were the maximum value of *w* in the daytime and the associated temperature and *RH*. The second variable would be the *RH* at 6:00 A.M.

Raw sensor data were extracted from the daily results over a period of a month and three months, and the two trends were duly plotted. The outcome showed that the daily maximum '*w*' in the panels had a random fluctuating pattern. However, the rising daily maximum '*w*' tendency for Panel 3, Panel 4, and Panel 9 was clearly observed. Examination of the 6:00 A.M. *RH* plots for the above three panels showed that Panel 4 and Panel 9 had a rising water content while Panel 3 had 60–80% *RH*. Visual inspection carried out on Panels 4 and 9 later confirmed that water had actually gotten into both the panels, which explains the rising tendency. *RH* in the other panels all fell below 65% at 6:00 A.M., except for Panel 2, which was below 85%.

Table 7.3 provides a summary of the average daily maximum '*w*' and *RH* at 6:00 A.M. The evaluation of the monitored panel results had to be based on the actual condition of the panels. High *RH* would substantially increase the possibility of condensation. However, this might not give us any clue for the test panels' corrosion condition because any water leakage into the panel would not immediately read a 100% *RH* at the probe tip, if the probe was located at some distance from the water. The panels were open systems,

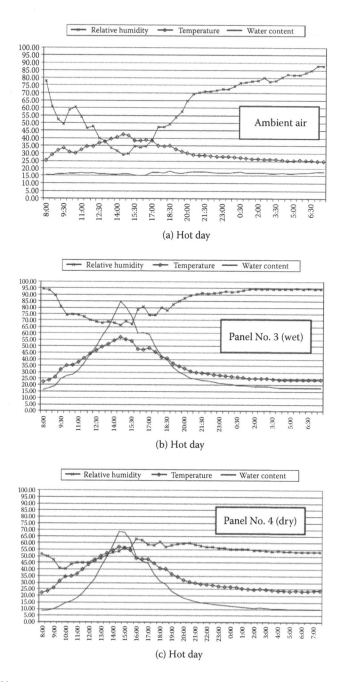

(a) Hot day

(b) Hot day

(c) Hot day

FIGURE 7.11
Typical daily monitoring results for a hot day (a), (b), (c); and a cold rainy day (d), (e), (f).

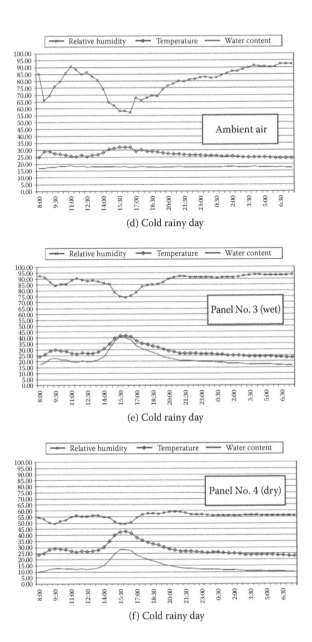

(d) Cold rainy day

(e) Cold rainy day

(f) Cold rainy day

FIGURE 7.11 (*Continued*)

where limited free air movement was piped through the hoses. Any water present resulting from a leak on a panel would be quickly absorbed by the air as well as the insulation materials. If the leak stopped, that little amount of water would be totally absorbed or evaporated into the air inside the panel, perhaps raising the *RH* level for a short while and eventually drying out

TABLE 7.3

Test Panel Results

Panel	Average Daily Maximum 'w' in the Daytime	Average *RH* at 6:00 A.M.
1	37.84	47.08
2	38.93	66.16
3	51.33	67.08
4	56.81	70.29
5	35.92	49.63
6	39.10	56.61
7	39.39	50.08
8	40.96	52.43
9	60.62	81.51
10	36.68	46.51
Ambient 11	18.80	88.56
Ambient 12	19.04	87.97

completely. On the other hand, if water continued to leak, the *RH* level would be maintained at a high level when the drying rate of the water was the same as the water leaking rate. But if the rate of leaking exceeded the rate of drying, there would be excess water and 100% *RH* saturation would show on the probe tip and across the whole panel too. Therefore, a three-month rising *RH* distribution would be a better indicator of water seepage. In the final analysis, the extent of the Z-spacer CUI in the panels (because of the presence of water) with respect to *RH* had to be further confirmed by visual inspection.

A visual inspection was then carried out on the 10 panels. Table 7.4 shows the average *RH* at 6:00 A.M. and the visual inspection results after six months. The extent of corrosion was estimated as the percentage of rust coverage area over the total upward facing surface area of the three Z-spacers in each panel. Clearly, severe corrosion and increasing *RH* were observed in both Panels 3 and 4. As for Panel 9, severe corrosion of 40% was already detected at the end of the one to three month period, when the average *RH* was 81.51%.

In conclusion, the average *RH* at 6:00 A.M. was a good indicator of corrosion. An average *RH* of 70% and below would not cause significant CUI. Hence, this was the criterion for an acceptable dry panel condition. This means if the *RH* was reduced gradually over the six-month period, the corrosion rate would controlled to an acceptable level.

As mentioned earlier, water content *w* does provide some information about the amount of water trapped in the cement boards. However, it is harder to conclude whether daytime *w* indicates a dry condition or not. A further investigation revealed that *w* has to be correlated to both temperature and *RH* to derive the optimum dry condition. This is shown in Figure 7.12. Maximum daily *w* rates were plotted against associated temperature rates for the test panels that were shown to be dry. An inspection of the plot shows

TABLE 7.4

Average *RH* at 6:00 A.M. and Visual Inspection Results

Panel	Average of 6:00 A.M. *RH* over Months 1–3	Average of 6:00 A.M. *RH* over Months 4–6	Results of Visual Inspection	Remarks
1	47.08	41.64	Very minor corrosion, ≈ 0%	Glass wool and cement board were dry.
2	66.16	66.8	Minor corrosion, 6.3%	Glass wool and cement board were dry.
3	67.08	78.00	Severe corrosion, 35%	Glass wool and cement board were damp.
4	70.29	72.56	Severe corrosion, 43%	Progressive corrosion observed. Glass wool and cement board were wet.
5	49.63	46.87	No sign of corrosion	Glass wool and cement board were dry.
6	56.61	53.09	Minor corrosion, 2%	Glass wool and cement board were dry.
7	50.08	65.54	Minor corrosion, 2%	Progressive corrosion observed.
8	52.43	47.12	No further detectable corrosion	Was wet once but current results showed dry condition. Glass wool and cement board were dry.
9	81.51	70.15	Severe corrosion, 40%	Progressive corrosion observed. Glass wool and cement board were wet.
10	46.51	39.14	No sign of corrosion	Glass wool and cement board were dry.

that all points fell within a band, delineated by an upper and a lower limit boundary. Results for dry bays with no ambient air entry should generally fall below the upper limit. Careful examination revealed that the upper limit is actually the 60% *RH* curve in a psychometric chart. Therefore, for maximum *w*, a maximum *RH* for dry condition was computed, the figure thus obtained was 60%. A curve for 100% *RH* was also included that represents the saturated conditions. Theoretically, this number cannot exceed 100%.

An important note about Figure 7.12 is that it is pertinent to evaluate only the test panels with little ambient air entry. The test panel conditions here are similar to that of the bays at the centre of the roof of the actual building investigated in this study. However, this condition differed in the bays at the edge of the roof where there was a fair amount of ambient air entry. Ambient air with high water content could add to the amount present in the roof, and increase the value of *w* and thus *RH*. It is not possible to theoretically separate the different sources of water, and since it is also not possible to simulate such effects accurately in the test panels, an accurate upper limit boundary

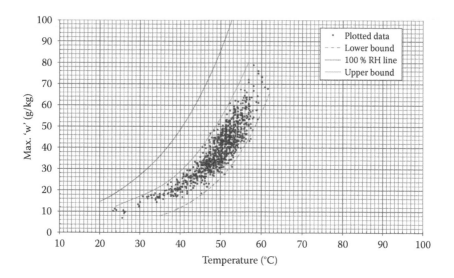

FIGURE 7.12
Plot of daily maximum water content and the associated temperature for dry panels.

cannot be obtained for the edge bays. For edge bays, it was suggested that the dry condition criteria should be based on the 6:00 A.M. maximum daily trend for w and RH. In brief, the following criteria were established that define the acceptable atmospheric conditions for the roof bays:

- Plots for the maximum daily w versus T: All points should fall below the upper boundary (not valid for edge bays).
- Trend for the maximum daily w: Values of w should not show a steadily rising trend.
- Trend for RH at 6:00 A.M.: Values of RH should not show a steadily rising trend and not exceed 70% for extended periods of time.

7.2.7.1 Criteria or Conditions Statements

1. Glass wool and cement board

   The presence of moisture in the glass wool and cement board would affect the environment RH in the roof space, so this factor had to be taken into account. Tests on unused glass wool kept in the laboratory gave a moisture content level that ranged from 0.29 to 0.43%. Experience in the roof showed that wool moisture contents of up to 3% did not cause fast corrosion. Hence, 3% is the criterion score for safe moisture in glass wool. Tests on unused cement board stored in the laboratory gave moisture contents ranging from 8.86 to 10.13%. To allow for experimental errors, 10.5% is the criterion score for safe moisture in the 3rd layer cement board.

2. Allowable increase in brown rust area at the corroded Z-spacers

Previous inspections of the Z-spacers in all the 30 bays showed that they were corroded to various extents. The base steel was covered with a layer of brown rust and if no further deterioration was found, the situation is considered to be acceptably dry. In addition, during the initial inspection, cuts were inadvertently made on the spacers, exposing bare steel. If no brown rust developed on the clean metal up to the time of the final inspection 13 months later, this would prove that the atmosphere there was relatively dry. The criterion statement was no increase in rust by area and size. Some rust confined to a small area and some new rust forming on recently exposed steel surfaces was quite reasonable.

7.2.7.2 Results from the Roof Bays

The characteristics observed in the test panels are also true for the roof space in the actual building. Small airflow is possible in the gaps between the 3rd layer cement board and the roof cladding that determines the moisture content in the roof. Under windy conditions, high- and low-pressure pockets are created. Air can move quite freely throughout the roof and building, and from the outside. Inside the building the situation is different; air can only travel across the vents at the keel sides that are located within the glazed walls. Note that the probe casings are not airtight so some air can penetrate.

In summary, air from the outside and inside of the building is confined to the gap in the roof between the 3rd layer cement board and the cladding. Therefore, it has no contact with the Z-spacers and will not cause them to corrode. On a cold and humid night, however, if sufficient ambient air gets in, condensation can occur under the roof cladding or on the top of the cement board. Condensates usually appear in the form of microfilm or micro droplets of water and it can then land on the Z-spacers. Nevertheless, condensates will disappear in the daytime so they pose little danger to the corrosion problem. Any ambient air is easily detected by the probes because their measuring tips are positioned between the 3rd layer cement board and the roof cladding. In contrast, relative humidity has a great effect on corrosion. Because of the psychometric properties of air, high RH may cause water condensation on the Z-spacer if the nighttime temperature difference between the roof atmosphere and the Z-spacer surface equals to or exceeds the value stated in Table 7.5.

For the roof space, initial experience suggests that monitored RH readings predominantly below about 75% would not cause fast corrosion. That corresponds to the 4.6°C difference between the roof space and the Z-spacer for condensation to occur, which is quite unlikely at nighttime. During the day, condensation is impossible because the average daytime temperature is 45°C and 60% RH up in the roof. A temperature difference of 13°C is required for condensation to settle on the Z-spacers, which is doubtful.

TABLE 7.5

Condensation Conditions at Night

RH (%) at 25°C at Night	Temperature Difference Causing Condensation (°C)
100	0
90	1.8
80	3.6
75	4.6
70	5.9
60	8.2

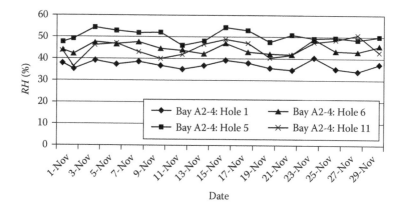

FIGURE 7.13
Monthly trend of *RH* at 6:00 A.M. for bay A2-4.

Figure 7.13 displays the typical monthly trends for bay A2-4, which shows dry conditions. Figure 7.14 presents the data for bay D12-2 which shows leaks developing in Hole 5. It also shows that *RH* values at 6:00 A.M. were steadily increasing in H5 (cf. the criterion developed for the test panels). In addition, slightly higher *RH* values were recorded in the daytime. These are indications of higher moisture content than normal. Note that an increase in *RH* is the first sign of a leak, though care must be taken to distinguish it from the ambient effect, which is usually temporary.

7.2.8 Corrosion Models

Corrosion found in the bare steel Z-spacer coupons was fairly uniform and no pitting was observed. Figure 7.15 shows the appearance of an exposed coupon. The corrosion rate was determined using the weight loss method and the results would not be underestimated since there was no pitting.

Test results showed that the three most severe corrosions occurred at A13 H6 (Bay A13, Hole 6), D122 H5, and D123 H5; in the magnitude of 0.009 mm/year.

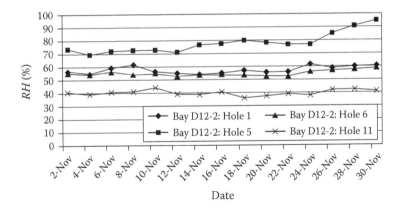

FIGURE 7.14
Monthly trend of *RH* at 6:00 A.M. for bay D12-2.

FIGURE 7.15
Uniform corrosion observed on bare steel Z-spacer coupons exposed at the roof.

The least corroded bay was at C24 at a rate of 0.00000816 mm/year. The mean value for corrosion in the roof was 0.000416 mm/year and the standard deviation was 0.00123 mm/year. Corrosion losses in all the galvanized steel Z-spacers as well as the flat sheet galvanized steel coupons were negligible. Visual examination also detected no further corrosion in the corroded galvanized steel coupons.

The corrosion rate of bare steel is described in the general kinetic relationship, Equation (7.1). Based on the calculations by the contractor who was responsible for the construction of the roof, a metal thickness of 0.7 mm would be sufficient for structural purposes. Since the Z-spacer is 1.6 mm thick and the warranty period is 50 years, it follows that the maximum

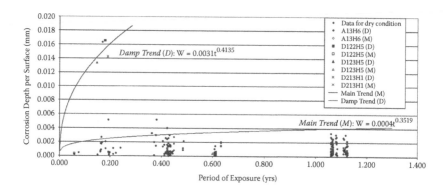

FIGURE 7.16
Corrosion depths of the bare steel coupons exposed in the roof space.

corrosion depth permissible per surface for 50 years would be 0.5(1.6–0.7)
= 0.45 mm. Therefore, the predicted corrosion depth per Z-spacer surface in
the roof should not exceed 0.45 mm in 50 years. Figure 7.16 shows the results
for the corrosion depth of exposed bare steel in the roof space within the
studied period. All points (except for four cases), fall into a zone that can be
delineated by the upper boundary line stated as:

$$\Delta W = 0.0004 \times t^{0.3519} \tag{7.2}$$

where W is the corrosion depth per surface (mm) and t is the period of
exposure (years). This is called the *main trend line* and represents the maximum corrosion rates for the majority of the exposed bare steel coupons.
The coefficient of determination of this line (R^2) is 0.8599 showing a fairly
good correlation. The four exceptional cases can be fitted into another line
given by:

$$\Delta W = 0.0031 \times t^{0.4135} \tag{7.3}$$

This line is called the *damp trend line* and represents the corrosion rate for
bare steel specimens under damp conditions. This happened because the
condition in the respective locations was not fully dry during the initial
stage of this study. Successive coupons from three of the four locations fall
within the main trend boundary, suggesting that they dried out afterward.
The two lines could be extended to 50 years and the following is predicted
(see Figure 7.17).

The corrosion depth per surface for the main trend is 0.00158 mm and the
remaining thickness of the bare steel coupon is 1.597 mm. The damp trends
are 0.0156 mm and 1.569 mm, respectively. This indicates that the corrosion
effect is negligible as the maximum corrosion depth per surface is estimated

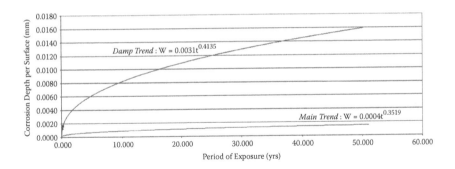

FIGURE 7.17
Prediction of the corrosion depth of bare steel Z-spacers in the roof.

to be 0.45 mm. The second corrosion model is based on dose-response functions [15–18], which relates the weight loss to the meteorological parameters (temperature and relative humidity) and air pollutants. The dispersed points in the main trend zone (Figure 7.16) suggest that besides time, there must be other variables affecting the depth of corrosion. Generally, corrosion is known to be influenced by RH, T, and air contaminants such as sulfur and chloride. This study is concerned with the depth of corrosion penetration of a metal surface. It predicts the surface thickness that is consumed by corrosion. The influence of air pollutants is not monitored in this study because of resource constraints. It is assumed that the stainless steel cladding in the roof serves as a weather barrier to shelter from rain, and thus bar any potential air pollutant entering the building. This eliminates the contribution of any possible air pollutants on the Z-spacer corrosion process. Despite that, EDS (energy-dispersive X-ray spectroscopy) analysis was performed on the rust found on the Z-spacer to check for pollutants indirectly. Sulphur was not significant, and chloride was found to be only 0.08% of the total rust weight. The following mathematical model is then proposed:

$$\ln(W) = b_0 + b_1 \ln(t) + b_2 RH_{av} + b_3 T_{av} \qquad (7.4)$$

where W is the corrosion depth per surface, t is time (year), RH_{av} and T_{av} are the average relative humidity and temperature over the coupon exposed period. Monitoring results of RH and T in the roof space were used in this analysis. However, there was very little information for extremely high RH conditions in the roof corrosion data. The roof under service should not have such a high RH. When high RH is detected, as in a case of leakage, the problem is immediately addressed. As such, a high-RH condition was virtually non-existent so no relevant corrosion data was collected. In this case, corrosion data points from the simulation panels were added to the selection in order to provide data for an extremely high-RH condition. The addition of panel data points is considered valid here assuming that the corrosion behaviors of the panel and roof space are similar. Parameters b_0, b_1, b_2, and b_3 can be determined by multiple regression analysis:

$$\ln(W) = 0.587\ln(t) + 0.0805RH_{av} + 0.0413T_{av} - 14.059 \qquad (7.5)$$

In this analysis, sample $N = 51$ and $R^2 = 0.48$.

In the main trend line shown in Figure 7.16, $RH_{av} = 59.5\%$, $T_{av} = 32.1°C$. Therefore the estimated corrosion depth per surface after 50 years is 0.00352 mm and the remaining thickness of the bare steel coupon is 1.593 mm. For the damp trend line, they are 0.0511 mm and 1.59 mm, respectively, with $RH_{av} = 64.7\%$, $T_{av} = 31°C$.

Equation (7.5) shows very good approximation to Vernon's curve [19], as is indicated in Figure 7.18. Substituting $RH_{av} = 64.7\%$, $T_{av} = 31°C$ using Equation (7.5) yields a corrosion depth of 0.063 mm after 50 years, which is only a 7% deviation from the value predicted in Vernon's curve B [19] extrapolation method. In fact, Equation (7.5) agrees well with the Vernon extrapolation curve, ranging from 40 to 80 years. Thus, it is very reliable to predict the life of the Z-spacers using Equation (7.5).

The environmental condition where the Z-spacer is exposed is free from airborne contaminants such as sulphate and chloride, and air particles and humidity are the only possible factors to promote further corrosion development. Vernon's curve was obtained from the corrosion results in the presence of airborne particulates and controlled RH. The fact that Equation (7.5) correlates well with Vernon's curve supports the viewpoint above and implies that the condition of the roof space is very similar to that in Vernon's experiment.

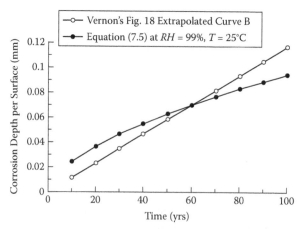

FIGURE 7.18
Comparison of Equation (7.5) and Vernon's extrapolated curve B.

The minor differences might be due to the effect of airborne particulates that were ignored in this study. In addition to temperature, humidity is the most likely main influence of roof space corrosion.

Both of the corrosion models showed a very low corrosion attack after 50 years, thus confirming the criteria established for acceptable dry conditions. Recall that the corrosion models are derived from data obtained for the three satisfactory dry condition criteria. These criteria were applied to roof bays to verify whether the roof condition is suitably dry. Comparison of *RH*, *T*, and *w* were carried out based on the first three- and last three-month monitoring trends (the total study period was one year). Results revealed that all 24 non-edge bays in the study showed signs of drying out as the maximum *w* and *RH* values at 6:00 A.M. had decreased slowly during the monitoring period. Meanwhile, all six edge bays in the study met the criteria of monitored *RH*, *T*, and *w* for stable and dry bays. From the selected bays, 398 glass wool and 291 cement board samples were taken. Moisture content tests were conducted on all the samples. During the initial tests (months 1–3), a large number of glass wool samples were found to have an average moisture content of 3.34%, in excess of the 3% criterion recommendation. However, in the final inspection (month 6) the moisture content of all the samples were actually very much lower than 3%, averaging only 0.445%. As for the cement board samples, 12 samples had moisture content higher than the 10.5% criterion level during the initial inspection period, but they all came in well within the criterion during the final inspection. The average moisture content was a low 6.9%. To conclude, all the roof bays fulfilled the three criteria requirements. That is to say they are confirmed to be adequately dry.

7.2.9 Conclusion

This case presented results of a study on the corrosion behaviors of metal-based roofing materials in a very large building. Severe corrosion under insulation on the steel structure was observed and the problem was rectified by blowing hot and dehumidified air to dry the roof. Three criteria were established, the application of which would ensure that the roof bays were dry. The criteria were: (1) the maximum daily water content should fall below the upper boundary (except for edge bays); (2) the maximum daily water content should not show a steadily rising trend; and (3) values of the relative humidity should not show a steadily rising trend and not exceed 70% for an extended period of time.

Two corrosion models were then developed, which defined the relationship between corrosion, relative humidity, temperature, and time. These models predicted a corrosion depth of the Z-spacer is acceptable for 50 years. It was deemed that under the environment pertaining within the roof space, bare steel and galvanized steel corrosion in the roof spaces of the large building would be negligible for the next 50 years.

7.3 Case Study 2: Condition Assessment and Monitoring of a College Building

7.3.1 Historical Background

The college building in this case is a fully residential school complex occupying a land area of about 50 acres. Approximately three years after it was built, some cracks appeared in the brick walls and other structural elements in the Administration and Academic Block. So the project contractor commissioned an engineering consultant to investigate the problem. Rectification work on the cracks was then carried out by the contractor himself after the study confirmed that the foundation and cracks had stabilized. However, after about two years, the cracks reappeared on the same building. Since the problem of crack propagations and foundation were not so obvious at that point in time, no appropriate action was undertaken. Four years later, two tubewells approximately 100 m from the Administration and Academic Block were constructed to obtain water from the ground. Since then, the existing cracks had accelerated in size and rapidity. Subsequently, new cracks appeared at another location; the door and window frames became distorted, the glass window panels broke, and the floor slabs subsided. The building was obviously not safe and the school engaged a consultant company to perform a condition assessment and monitoring program.

7.3.2 Scope of Commission Works

The scope of the commission work involved the following tasks:

a. A comprehensive visual inspection and distress mapping assignment for all the structural and non-structural elements of the building block.

b. A monitoring program, comprising the installation of ground and building settlement markers and Demec crack width monitoring stations at selected locations.

c. A soil and foundation investigation work comprising a rotary wash boring of soil, foundation/trial pit excavation, and Mackintosh probe test.

d. Material investigation work consisting of rebar scanning using a Ferroscan machine and the extraction of concrete core samples at selected locations.

e. Laboratory tests on extracted soil and concrete samples for items (b) and (c) above.

f. An interpretation of all test results.

g. A condition assessment of the building based on all the available test results.

h. Conceptual recommendations and cost estimates for rehabilitation work.

7.3.3 Description of the Administration and Academic Building

The Administration and Academic Block is one of many buildings in the college complex. It is utilized as a centre for administration and academic affairs. It also houses a number of classrooms, some laboratories as well as a library. The two-story block was built according to the Conventional Reinforced Concrete (R.C.) construction method. The building has an overall dimension of 70.2 m (w) × 140.4 m (l). Conventional flat slab construction was used in the first floor slab at gridline A-J/4-7 and D-G/16-18 (see the layout plan shown in Figure 7.19). The reinforced concrete slabs rested on main and secondary reinforced concrete beams spanning some reinforced concrete columns. The perimeter and in-filled brick walls were constructed

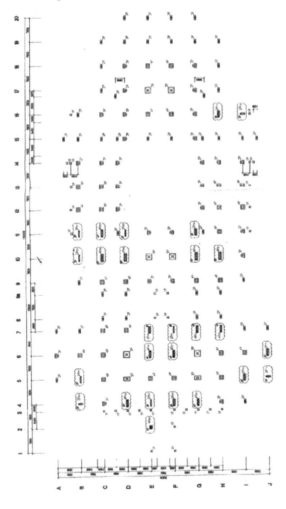

FIGURE 7.19
Building a layout plan.

from column to column forming the perimeter of the building block. The walls were erected with common bricks using cement-sand plaster.

7.3.4 Comprehensive Visual Inspections and Distress Mappings

Comprehensive visual inspections and distress mappings were conducted to map out all forms of distresses on the structural and non-structural elements of the building block throughout the entire duration of the investigation work. All the distress signals were systematically observed and recorded during the visual inspections. Generally, no significant cracks were seen on the structural and non-structural elements of the building block. However, the ground floor slabs had subsided damaging the tiles at gridline E-F/4-6 and A-J/6-7. Some bulging or dislodged floor tiles were also observed on the top of the first floor slab at gridline F-G/10-11. There was a reasonably large gap at the joints between the first floor column and the brick wall at gridline F/16. Close observations revealed that the suspended ceiling above this area was damaged due to lateral movement of the brick wall. Cracks with signs of water leakage were seen at the soffit of the perimeter R.C. gutter as well as on the pitch roof tiles at several locations. The majority of these leaks were still active. Checks on the surrounding perimeter drain indicated several damaged portions of the drain resulting in the water draining into the ground when it rains. Figure 7.20 shows some photographs of the distresses found on the building elements.

7.3.5 Instrumentation for Monitoring Work

The following four instruments were installed for condition monitoring:

i. Temporary Bench Mark (TBM)

Temporary Bench Mark (TBM) is a permanent reference station used as a reference for leveling measurements to other settlement markers and should be established on stable ground or a structure. In this instance, there were no visible or available original bench marks around the site. So for monitoring purposes, a temporary bench mark was established by putting a nail on the top of a concrete drain sump near the Guard House.

ii. Ground Settlement Markers

These markers are constructed on the ground to measure the ground levels for monitoring the ground settlement. A rounded tip steel rod of 20 mm in diameter and approximately 600 mm long is driven into the ground leaving 50 mm of the top protruding out of the ground. The surrounding of the rod is then covered with non-shrink grout.

iii. Column Settlement Markers

These markers are installed on selected columns to monitor possible column settlement. In this case, a hole of 25 mm in diameter and

FIGURE 7.20
Photographs of some distresses observed on the building elements.

70 mm depth was drilled into the lower part of the column where a stainless steel socket was placed. During the leveling work, a removable ball-end leveling plug was screwed into the socket on which a flat-based staff was held for reading the precise levels.

iv. Demec Crack Width Gauges

These gauges are installed over an existing crack width to monitor any possible propagation or changes in the crack width with time (Figure 7.21). Two steel studs of 3 mm diameter are epoxied onto the wall on each side of a crack line. During crack monitoring, the removable mechanical gauge is mounted onto the studs and a reading is obtained from the dial.

7.3.6 Soil and Foundation Investigation Work

In order to obtain information about the ground properties, much soil investigation work was carried out. Eight rotary wash boreholes were used to determine the subsoil profile and general soil properties as demonstrated in Figure 7.22. These boreholes were conducted by using a multi-speed hydraulic feed rotary-boring machine with a mast. The technique for advancing the

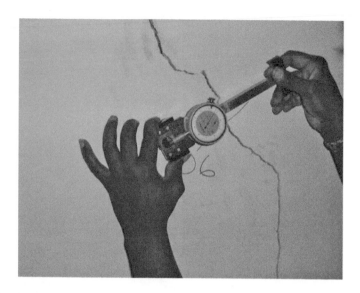

FIGURE 7.21
Crack width monitoring using a Demec gauge.

FIGURE 7.22
Rotary wash boring to establish the soil properties.

borehole involved wash boring or the cutting and flushing action of water that was pumped out at high pressures through the casings or rods drilled into the bottom of the hole. The wash bore was periodically stopped at specific depths for conducting the standard penetration tests and for collecting the disturbed soil samples. Change of soil strata during the boring operation was recorded using the flushed materials and soil samples.

In the meantime, the standard penetration tests (SPT) were performed in accordance with BS 5930: 1999 [20], using an open-ended longitudinally split spoon sampler with a cutting shoe at the lower end and a coupling at the upper end connecting the sampler to the AW drill rod. The sampler was driven into the soil by a 63.5 kg automatic trip hammer free-falling through a height of 760 mm onto an anvil. A number of blows were required to effect a 300 mm penetration (test drive) below an initial penetration of 150 mm (seating drive) through the soil at the bottom of the borehole. The penetration resistance is called the N-value. The SPTs were conducted at a depth of 1.0 m for the first 10 m of drilling and subsequently at 1.5 m intervals for the rest of the drilling. The N-value and the length of the disturbed samples recovered from the sampler were indicated in the borehole logs. The relative density of the non-cohesive soils and the consistency of the cohesive soils were then defined, according to the penetration resistance, N-values. The relationship between the N-values and relative density and soil consistency were derived from Terzaghi, Peck, and Mesri [21]. To monitor the groundwater level, standpipes were installed onto each borehole to study the groundwater table. Readings were then taken every day throughout the duration of the soil investigation work.

In addition, a Mackintosh probe test (Figure 7.23) was used to get an idea about the soil stiffness at different depths below the foundation. The test is often employed as a random check for the degree of compaction on selected areas of soil. The termination criteria for a Mackintosh probe test is 400 blows/foot or 12.5 m maximum, whichever is achieved first. Some selected foundations were deliberately exposed to determine the foundation system configuration adopted in the construction of the building as well as

FIGURE 7.23
A Mackintosh probe test in progress to investigate soil stiffness at different depths below the foundation.

to cross check them against the available contract drawings. Soil tube and bulk samples were collected to ascertain the soil *in situ* density and for the laboratory compaction test.

7.3.7 Material Testing

Nominal on-site material testing was done randomly at selected locations to check the concrete strength and durability qualities of the reinforcements as well as the material condition of the original structure.

7.3.8 Electronic Rebar Scanning

The Ferroscan FS-10 machine for electronic rebar scanning was used to locate the exact location of the steel reinforcement bars in the concrete members prior to the concrete core extraction. It is also used to locate the steel reinforcement bars in the concrete members and to estimate the concrete cover for the steel reinforcement bars. Adequate concrete cover is necessary to ensure that the steel bars are properly protected against corrosion. Take note however that excessive concrete cover can also give rise to other problems.

7.3.9 Concrete Core Testing

Immediately after extraction, the concrete core samples were subjected to carbonation testing to ascertain the depth of carbonation in the concrete. Carbonation occurs from atmospheric carbon dioxide penetration. Carbon dioxide plus water becomes carbonic acid which will neutralize the alkaline in the cement matrix. Carbonation proceeds from the surface inwards. The rate of penetration depends on factors such as age, relative humidity, and concrete permeability. Concrete has a high pH of 12.5–13.0. As the layer of concrete closest to the surface becomes carbonated, its alkaline decreases and may become insufficient to protect the steel reinforcement bars against corrosion. If the carbonation depth goes beyond the steel reinforcement bars then protection afforded by the alkaline concrete cover is lost. This may then trigger off reinforcement corrosion if the right environmental conditions are present. Once reinforcement corrosion sets in it will proceed rapidly leading to early loss of structural strength.

According to BS 1881: Part 114: 1983 [22], a good concrete mix proportion and adequate concrete compaction produces dense concrete that meets the strength requirement and provides durability protection at the same time. Generally speaking, a reasonably well-compacted concrete has a density of not less than 2200 kg/m^3. Concrete compressive strength is an important property of concrete. This strength will determine the load-carrying capacity of the structural component. The concrete density and compressive strength tests for the extracted core samples were done in the laboratory.

7.3.10 Discussion of the Results

7.3.10.1 The Monitoring Results

A total of 10 ground settlement markers, 37 column settlement markers, and 29 Demec crack width monitoring stations were installed to monitor the movement behavior around the building block. Monitoring stations were set up on selected columns in the building that had severe visible distresses. Places with fewer distress marks were also monitored for comparison purposes. During the 20-day monitoring period, a total of 15 monitoring trips were completed. Generally, all the ground settlement markers showed some fluctuation and heaving throughout the monitoring duration. For column settlement markers, 37 of them were divided into two groups on the basis of their severe distress signals.

   i. *Group 1*: CS01 to CS19 and CS31 and CS37
   ii. *Group 2*: CS20 to CS30

For *Group 1*, markers CS01 to CS05 (minimum –0.4 mm to maximum –6.8 mm.); markers CS31 to CS373 (3 to –3 mm).
   Group 2, settlement activity in first two readings; heaving for rest of the monitoring period.
   All the Demec crack width monitoring stations except for D24 recorded stable crack propagation and crack width fluctuations ranging from 0.3 to –0.2 mm. A continuous crack in D24 had widened to 1.1 mm. However, these cracks were beginning to stabilize towards the end. Notwithstanding, the supervision duration should be extended in order to learn more about their behavioral patterns.

7.3.10.2 The Soil and Foundation Investigation

Eight rotary wash boreholes were examined (BH1–BH8), seven of them around the perimeter of the building block and one near the tubewell. The tasks encompassed obtaining disturbed and undisturbed soil samples for the laboratory tests, excavation in selected areas to study the foundation system, and extraction of soil samples for compaction testing in the laboratory. A Mackintosh probe test was staged at each of the excavation/trial pit locations and other selected areas surrounding the building block.
   The results showed that the soil was clayey silt material in all the eight boreholes. The depth for SPT values greater than 50 was between 9.0 and 17.5 m. The topsoil was 6–10 m thick, very soft to firm clayey silt. Underneath the topsoil was very stiff clayey silt, and below this layer still was competent hard stratum (SPT ≥ 50). The two bottom layers were 10 and 15 mm deep. The soil liquid limit ranged between 41 and 53%, the plasticity index between

10 and 28%, and the groundwater level between 0.7 and 7.0 m. The filled materials consisted of clayey silt with traces of sand and gravel, and they contained an SPT value of 1 to 39 blows/0.3 m (average = 20) and moisture content 23 to 36% (average = 30%). Finally, a total of seven trial pits were excavated to get a picture of the actual foundation lay of the building. The study found that the system complied with the building specifications.

7.3.10.3 Material Testing—Rebar Scanning

The Ferroscan machine scanned a total of 87 rebar on selected R.C. beams, columns, slabs, and drop panels. The rebar scans indicated the following three results. The figures (%) refer to the percentage of areas scanned. The brackets enclose the values/range (mm) of concrete cover provision (consistent, inconsistent, excessive, or average).

    i. R.C. beams
- All areas (100%)—Generally inconsistent
- 57%—Excessive (35–89 mm)
- 43%—Consistent average (14 and 32 mm)

    ii. R.C. columns
- 80%—Excessive (35–98 mm)
- 20%—Consistent average (18 to 34 mm)

    iii. R.C. slabs and drop panel
- Two scans on two selected slab panels—Average (16 and 19 mm)
- One selected first floor slab drop panel—Average (37 mm)

7.3.10.4 Material Testing of the Core Samples

Some concrete core samples taken from the site were subjected to a carbonation test. The carbonation depth was determined by applying phenolphthalein solution on the samples. The depths, which relate to the age and durability properties of the concrete, ranged from almost negligible to about maximum 20 mm. The findings revealed that carbonation had not reached the rebar zone of the structural elements, suggesting that no rebar corrosion-related problems would occur in the near future. The test found the presence of plastering and skim coats which also provided the structural components some form of protection. Results of the statistical analysis revealed that 95% of the maximum carbonation depth fell within the range of 0 to 14 mm.

Separately, another test was performed to gauge the density of the concrete material. The test results showed them to be generally above 2200 kg/m³. This meant that the concrete used in the building was very well compacted or dense. Overall, 95% of the concrete density was above 2230 kg/m³. That was a good sign. To estimate the *in situ* concrete strength, 26 concrete core samples underwent the compressive strength test in the laboratory. Below are the findings for the compressive strength of the columns, beams, slabs, and pile caps:

i. Columns
- Six core samples from the ground and first floor columns (17.5–28.0 N/mm², mean 22.5 N/mm²)

ii. Beams
- Three core samples from selected ground beams (26.5–31.0 N/mm², mean 28.8 N/mm²)
- Five core samples from selected first floor beams (18.0–29.5 N/mm²); core C14 (10.5 N/mm²)
- Two cores from selected roof beams (19.5 and 10.0 N/mm²)

iii. Slabs
- Three core samples from selected first floor slab (20.5–25.5 N/mm², mean 22.5 N/mm²)

iv. Pile caps
- Six core samples from exposed pile caps in the trial pit area (21.5–28.0 N/mm², mean 24.2 N/mm²)

The results of a statistical analysis of overall concrete compressive strength revealed that 95% exceeded 14 N/mm². Various compressive strengths were also shown, which means the concrete varied in quality. This must be limited to a small area because only two cores had an exceptionally low concrete strength of 10.0 N/mm².

7.3.11 General Structural Inspection

A cursory inspection was made on a few first floor and roof beams to get an idea of the general layout provided in the contract drawings, crack history, previous repair work, fortification details, and so on. The structural capacity, maximum bending moment, and shear force of the existing structure could be estimated based on the reinforcements made. The existing structure's safety factor could be estimated based on these facts and its working load condition. The first floor beams labeled FB6, FB19, and FB22 were duly checked out. The results revealed that all of these beams would not require strengthening as the crack width and safety factor were satisfactory. However, the shear force at beam FB19 was weak (safety factor 0.89 and crack width 0.35 mm) so this section needed some reinforcement work done.

7.3.12 Condition Assessment: Geotechnical Aspects

The results obtained in the soil investigation and the general site survey of the Academic and Administration Block suggest that the building is located partly on excavated land and partly on filled land. The rear part of the building, between Gridlines 11 and 20, is sitting on excavated land. The front part, between Gridlines 1 and 11, is filled land which has a fill thickness up to about 10 m near Gridline 4.

7.3.12.1 Soil Investigation

A total of eight boreholes were drilled around the site. The subsoil generally consists of clayey silts. But boreholes BH-1, BH-2, BH-3, BH-7, and BH-8 have very soft to firm (SPT-N ≤ 4) clayey silts extending from the ground surface to a depth of 6 to 10 m. This layer of soft soil is most likely some sort of filler earth that was dumped here in the past. Otherwise, the rest of the building stands on competent hard stratum of SPT-N ≥ 50 between 10 and 15 m from the ground surface.

7.3.12.2 Underground Water

The groundwater levels in boreholes BH-1 to BH-8 varied between 0.7 and 7.0 m below the ground surface. The average depth of the groundwater is about 4.8 m.

7.3.12.3 Tap Groundwater from the Tubewells

Two tubewells (Table 7.6) were installed adjacent to the building to tap the groundwater.

7.3.12.4 Consolidation Settlement and Possible Causes

Consolidation settlement is the vertical displacement of the ground surface. This means that there is a volume change in the saturated cohesive soils because excess pore water pressure that occupies the void spaces or pores between the soil particles has become dissipated. The time-dependent rate of consolidation depends on the permeability of the subsoil. The time is relatively short with granular permeable soil, but longer with cohesive impermeable soil. The magnitude of consolidation settlement can be significant,

TABLE 7.6

Details of the Tubewells

Particular	Tubewell TW/01	Tubewell TW/02
Distance from centre of building	About 123 m	About 99 m
Pump installation depth	68.0 m	41.0 m
Well depth	85.0 m	96.0 m
Approximate well yield	10.45 m³/hr (251 m³/day)	9.00 m³/hr (216 m³/day)

particularly in a compressible soft soil layer. There are a few possible causes of consolidation settlements, as discussed below:

i. Lowering of groundwater level

It was not realized that settlements could be caused by a lowering of the groundwater level until Terzaghi's investigations in the 1920s when the concept of effective stress was proposed. Effective stress σ' is equal to total stress σ less the pore water pressure u, as expressed by this equation for saturated soils.

$$\sigma' = \sigma - u \qquad (7.6)$$

Drawdown of the groundwater table can cause a reduction in the pore water pressure in the subsoil resulting in an increase in effective stress. This process leads to consolidation and settlement of the subsoil. A classic example is Mexico City where lowering of the groundwater level has caused vast areas of the land around the city to settle or subside. In 1948, parts of the city were settling at a rate of about 1 mm per day. Since measurements began in 1898, more than 6 m of subsidence has been recorded. In this particular city, consolidation settlement was possibly due to several factors including lowering of groundwater level beneath the building.

ii. Soft soil underfill or structure loads

Consolidation settlement exists if a structure or fill is placed (i.e., an additional load is imposed) over an existing layer of compressible soft soil.

iii. Improperly compacted fill materials

Consolidation settlements can also result from improperly compacted fill materials as they compress under their own weight or under the weight of further soils, surcharge, or structures placed over the fill area. If the top 6 to 10 m thick very soft to firm clayey silts encountered in boreholes BH-1, BH-2, BH-3, BH-7, and BH-8 are fill materials, then consolidation settlements exist.

7.3.12.5 Negative Skin Friction on Pile Shafts

Consolidation settlements of subsoil result in downward movements. If pile shafts are used for supporting structures in areas undergoing consolidation settlements, these downward movements will result in drag-down forces on the pile shafts, a process generally known as negative skin friction. Vesic [23] states that a relative downward movement of as little as 15 mm of the soil with respect to the pile may be sufficient to mobilize full negative skin friction.

The occurrence of negative skin friction on the pile shaft is similar to imposing an additional compressive load onto the existing pile. As a result, the existing pile may have failed to carry the load that it was designed to support within the allowable serviceability limits, thus resulting in excessive pile settlements. If the additional load to carry is too excessive, the existing pile or pile group can fail in its load-carrying capacity, either in structural capacity or geotechnical capacity. Unfortunately, it is often not easy to determine whether an individual pile or pile group have failed in its capacity or have merely experienced excessive settlements in a frame structure because the capability of load sharing and redistribution to other adjacent pile or pile group members in a frame system often makes the actual situation imperceptible.

7.3.12.5.1 *Deliberation of the Problems*

Generally, the problems arising at the building block are related to the following factors:

i. Consolidation settlements due to lowering of groundwater level. In fact, monitoring results from the ground settlement markers and column settlement markers show that the ground and columns (hence piles) are experiencing heaves, instead of settlements, during the recharging of the groundwater to its original level.

ii. Consolidation settlements due to improperly compacted fills. This problem may still remain but is most likely insignificant after about 10 years from completion of the building. Most of the consolidation settlements should have occurred during the past 10 years, and the rate of consolidation should continue to reduce with time.

iii. Excessive pile settlements and, perhaps, insufficient load-carrying capacities in individual piles or pile groups due to negative skin friction induced on the pile shafts. It should be noted that even if the induced negative skin friction has been fully removed from the pile, the pile settlement may not be fully recoverable (i.e., plastic settlement) and a reduction of soil strength to its residual strength may occur if the critical state strength of soil has been exceeded before.

iv. Unknown condition of the existing affected piles, that is, any pile damage caused by the negative skin frictions in the past?

v. Unknown actual loads imposed on piles after load sharing and redistribution in a frame structure due to differential settlements and performances of piles.

vi. Unknown actual pile length and number of the piles driven.

vii. Uncertainties of the previous foundation design as it was informed that cracks had been observed since completion of the building.

7.3.12.6 *Proposed Remedial Work*

When the ground settlement markers and column settlement markers were monitored for the last time, the results showed no more negative skin friction in the pile shafts. Nevertheless, the team still has to safeguard the long-term performance of the pile foundations. There are also many uncertainties and disadvantages to be considered as outlined in (iii) to (vii) above. In view of this, the following remedial work is proposed:

i. Ignore the working load capacity of the affected foundation piles for the underpinning design.
ii. Strengthen the affected pile foundations by underpinning with an adequate number of micropiles. A minimum of two micropiles are required for each affected pile group.
iii. Consider adding new pile loads in the existing and proposed new pile foundations.

Some old ones were destroyed by negative skin friction.
Details of the proposed micropiles:

- Diameter of micropile = 200 mm
- Estimated length of micropile = 20–30 m
- Working load of micropile = 50 tonnes
- Use minimum Grade 30 non-shrink cement grout

7.3.13 Condition Assessment: Structural Aspects

The occurrence of differential ground settlement had caused distortion and deflection on some structural elements of the building block as well as architectural damage, such as distortion of door frames, broken windows, and damaged ceiling boards. In spite of this, the overall structural integrity of the building block is deemed to be stable at this point of inspection. Structural cracks observed on several first floor beams and slabs should be reinstated and strengthened accordingly. The contract drawings claimed to have several first floor beams but they were missing. Hence, the first floor flat slabs have to be strengthened in order to make the structure more rigid and upright.

7.3.14 Recommendations

Results of the investigation show that the Academic and Administration Block of the college complex is situated partly on excavated land and partly on filled land. One of the causes of the structural and architectural damage to the building was due to very soft ground soil resulting from fill earthwork in the past. The average depth of the groundwater is roughly 4.8 m. Consolidation settlements may result from: (i) lowering of groundwater level; (ii) soft soils underfill or structure loads; and (iii) improperly compacted fill materials. Consolidation settlements of subsoil result in downward movement which induces negative skin frictions on the pile shafts, causing excessive pile settlements, bearing capacity failure, and/or pile damage. The behavior of the pile foundation is unpredictable once the earth settles or is settling down. Probably a combination of factors contributed to the pile failure that is not immediately obvious. The building on the whole is still strong and should not be demolished, as it is still fit for human occupation. On the basis of the visual inspections, monitor program, laboratory tests, and general supervision, the following recommendations are made for restoring the structural integrity as well as aesthetics of the Administration and Academic Block, for long-term use.

a. Foundation strengthening by underpinning method using micropiles.

b. Superstructure strengthening to enhance rigidity of the building.

c. Ground floor R.C. columns strengthening to eliminate distress.

d. Concrete repairs.

i. Strengthen the beams with low compressive strength using carbon fiber wrapping.

ii. Repair all non-structural cracks on walls, slabs, and beams using the routing method.

iii. Repair and waterproof leaks in R.C. gutter and roof.

iv. Repair concrete perimeter drain and correct invert levels.

e. Other rectification work.

i. Replace settled floor slab at main entrance.

ii. Reinstate and level damaged floor tiles within gridline F-G/10-1.

iii. Remove carpet over uneven floors and put in new tiles.

iv. Restore damaged doors, glass windows, and frames.

v. Apply new coat of paint for the whole building.

7.4 Conclusion

This chapter presented two case studies of failures encountered in large-scale infrastructures or constructions. The first case relates to the problem of corrosion under insulation (CUI) in a metal-based roof of a large building.

Water had entered the insulation zone of the roof during its construction. The problem is rectified by blowing hot and dehumidified air to dry out the roof. Condition assessment on the roof is sorely required but the inspection is to implement because the exposure would cover a very large space and area. In view of that, a more economical and efficient method was recommended. This entails formulation of some criteria for the maintenance of an acceptable dry condition mechanism in the roof section that would prevent immediate and long-term corrosion to the Z-spacers and other components. Based on the findings obtained in the simulation test panels, three criteria are established. The criteria were (1) the maximum daily water content should fall below the upper boundary (except for edge bays); (2) the maximum daily water content should not show a steadily rising trend; and (3) values of the relative humidity should not show a steadily rising trend and not exceed 70% for an extended period of time. Subsequently, all the roof bays are certified to be in a dry state or condition as defined by the three criteria. The corrosion models proposed also predicted that the corrosion depth of the roof materials which are made of bare steel and galvanised steel will stay well below the maximum corrosion depth in the next 50 years.

The second case study involves condition assessment and monitoring work conducted on a college building. The first task was to evaluate the structural integrity as well as the extent of settlement and crack propagation inside the building. A comprehensive visual inspection was carried out to observe and record the distresses. To do this several instruments were installed for condition monitoring such as Temporary Bench Mark (TBM), ground and column settlement markers, and Demec crack width gauges. Eight rotary wash boreholes were dug to test the subsoil profile and general soil properties. Material investigation work comprising rebar scanning using a Ferroscan machine and extraction of concrete core samples at selected locations were performed. Generally, the source of the problems were consolidation settlements, excessive pile settlements, and very possibly insufficient load-carrying capacities of the individual piles or pile groups due to negative skin friction induced on the pile shafts. On the basis of the tests performed, recommendations were made which would restore the overall integrity and aesthetics to its original condition. This work includes foundation and superstructure strengthening, concrete repairs, and other minor rectification work.

Problems and Answers

Problem 7.1

What are the characteristics of corrosion under insulation (CUI)?

Answer 7.1

CUI is characterized by localized corrosion and pit formation. It often occurs at the insulation interface of a metal surface, for example, in wool-insulated metal components. Practically all structural metals are susceptible.

Problem 7.2

What are the three important parameters to be monitored for establishing an acceptable dry condition for a roof bay?

Answer 7.2

Relative humidity, temperature, and water content. Note that the last parameter needs to be calculated from the first two.

Problem 7.3

The corrosion behavior of the galvanized steel Z-spacer used in the construction of a large building roof is governed by a dose-response function

$$\ln(W) = 0.587 \ln(t) + 0.0805 RH_{av} + 0.0413 T_{av} - 14.059$$

where W is the corrosion depth per surface (mm), t is time (year), RH_{av} (%), and T_{av} (°C) are the average relative humidity and temperature over the exposed period.

If the original thickness of the Z-spacer is 1.8 mm, estimate its remaining thickness after 50 years given $RH_{av} = 70\%$, $T_{av} = 32°C$.

Answer 7.3

$$\ln(W) = 0.587 \ln(t) + 0.0805 RH_{av} + 0.0413 T_{av} - 14.059$$

$$\text{Substitute } t = 50, \, RH_{av} = 70, \, T_{av} = 32$$

$$W = 0.00818 \text{ mm}$$

Hence the remaining thickness is 1.8–0.00818 = 1.79182 mm.

Problem 7.4

What are the common inspection procedures for conducting a condition assessment and monitoring work at buildings?

Answer 7.4

(a) Comprehensive visual inspection and mapping of distresses on all structural and non-structural elements of the building block;

(b) Installation of instruments for monitoring work such as ground and building settlement markers and Demec crack width monitoring stations;

(c) Soil and foundation investigation work comprising rotary wash boring of soil, foundation/trial pit excavation work, and Mackintosh probe test;

(d) Material investigation work comprising rebar scanning using a Ferroscan machine and extraction of concrete core samples;

(e) Laboratory tests on extracted soil and concrete samples;

(f) Interpretation of all test results and then carry out condition assessment based on the test results, and

(g) Conceptual recommendations and cost estimate for rehabilitation work.

Problem 7.5

Consolidation settlement is the vertical displacement of the ground surface corresponding to the volume change in saturated cohesive soils. What are the possible causes of consolidation settlements?

Answer 7.5

Possible causes of consolidation settlements:

- Lowering of groundwater level
- The structure or fill is placed (i.e., an additional load is imposed) over an existing layer of compressible soft soil
- Improperly compacted fill materials

References

General References

ASM Handbook, Vol. 13A, *Corrosion: Fundamentals, Testing, and Protection*, Ed., S.D. Cramer and B.S. Covino, Jr., ASM International, Ohio, 2003.

Specific References

1. Barnhart, J.M., The Function of Thermal Insulation, *Corrosion of Metals under Thermal Insulation*, ASTM STP 880, Ed., W.I. Pollock and J.M. Barnhart, Philadelphia: American Society of Testing and Materials, 1985.
2. Ashbaugh, W.G. and T.F. Laundrie, A Study of Corrosion of Steel under a Variety of Thermal Insulation Materials, *Corrosion of Metals under Thermal Insulation*, ASTM STP 880, Ed., W.I. Pollock and J.M. Barnhart, Philadelphia: American Society of Testing and Materials, 1985.
3. Dharma, A., W.G. Ashbaugh, R.D. Kane, N. McGowan, and B. Heimann, Measurement of Corrosion under Insulation and Effectiveness of Protective Coatings, Corrosion/97, Paper No. 266, Houston, TX: NACE International, p. 266.
4. Kumar, M.S., M. Sujata, M.A. Venkataswamy, and S.K. Bhaumik, Failure Analysis of a Stainless Steel Pipeline, *Eng. Fail. Anal.*, Vol. 15, No. 5, 2008, 497–504.
5. Elshawesh, F., A. El Houd, and O. El Raghai, Corrosion and Cracking under Insulation of Type 304 Stainless Steel at Ambient Temperature, *Corros. Eng. Sci. Technol.*, Vol. 38, No. 3, 2003, 239–240.
6. Delahunt, J.F., Proc. Exxon Research & Engineering Co., Internal Conference on Corrosion under Insulation, 1984, p. 554.
7. Martin-Regueira, Y., O. Ledea, F. Corvo, and C. Lariot, Corros, Indoor Atmospheric Corrosion of Copper and Steel under Heat Trap Conditions in Cuban Tropical Climate, *Corros. Eng. Sci. Technol.*, Vol. 46, 2011, 624–633.
8. Castano, J.G., C.A. Botero, A.H. Restrepo, E.A. Agudelo, E. Correa, and F. Echeverria, Atmospheric Corrosion of Carbon Steel in Colombia, *Corros. Sci.*, Vol. 52, 2010, 216–223.
9. Cole, I.S., W.D. Ganther, S.A. Furman, T.H. Muster, and A.K. Neufeld, Pitting of Zinc, Observations on Atmospheric Corrosion in Tropical Countries, *Corros. Sci.*, Vol. 52, 2010, 848–858.
10. Veleva, L., E. Merez, and M. Acosta, Zinc Precipitation-Runoff from Galvanized Steel in Humid Tropical Climate, *Corros. Eng. Sci. Technol.*, Vol. 46, 2010, 76–83.
11. Potgieter-Vermaak, S.S., A. Mmari, R. van Grieken, R.I. McCrindle, and J.H. Potgieter, Degradation of Galvanised Iron Roofing Material in Tanzania by Atmospheric Corrosion, *Corros. Eng. Sci. Technol.*, 2011, Vol. 46, 642–650.
12. British Standards Institution, British Standard. *Structural Use of Steelwork in Building Code of Practice for Design*, BS5950, Part 1, London: BSI, 2000.
13. Japanese Standards Association, *Japanese Industrial Standard for Hot-Dip Zinc-Coated Steel Sheets and Coils*, JIS G 3302–1994, Tokyo: Japanese Standards Association, 1994.
14. American Society for Testing and Materials, Standard Practice for Preparing, Cleaning and Evaluating Corrosion Test Specimens G1-90, *Annual Book of ASTM Standards*, Vol. 03.02, West Conshohocken, PA: ASTM, 1996, 9–15.
15. Mikhailov, A.A., M.N. Suloeva, and E.G. Vasilieva, Environmental Aspects of Atmospheric Corrosion, *Water, Air, Soil Pollut.*, Vol. 85, No. 4, 1995, 2673–2678.
16. Leuenberger-Minger, A.U., B. Buchmann, M. Faller, P. Richner, and M. Zöbeli, Dose-Response Functions for Weathering Steel, Copper and Zinc Obtained from a Four-Year Exposure Programme in Switzerland, *Corros. Sci.*, Vol. 44, 2002, 675–687.

17. Ferm, M., J. Watt, S. O'Hanlon, F. de Santis, and C. Varotsos, Deposition Measurement of Particulate Matter in Connection with Corrosion Studies, *Anal. Bioanal. Chem.*, Vol. 384, 2006, 1320–1330.

18. Kreislova, K., D. Knotkova, and L. Kopecky, Changes in Corrosion Rates in Atmospheres with Changing Corrosivity, *Corros. Eng. Sci. Technol.*, Vol. 44, 2009, 433–440.

19. Vernon, W.H.J., A Laboratory Study of Atmospheric Corrosion of Metals, *Trans. Faraday Soc.*, Vol. 31, 1935, 1678–1700.

20. British Standards Institution, *The Code of Practice for Site Investigations*, BS5930, London: BSI, 1999.

21. Terzaghi, K., R.B. Peck, and G. Mesri, *Soil Mechanics in Engineering Practice*, 3rd ed., New York: John Wiley & Sons, 1996.

22. British Standards Institution, *Testing Concrete. Methods for Determination of Density of Hardened Concrete*, BS1881–114, London: BSI, 1983.

23. Vesic, A.S., Design of Pile Foundations, National Cooperative Highway Research Program, *Synthesis of Highway Practice*, Vol. 42, Transportation Research Board, Washington, DC: National Research Council.

8

Radiation-Induced Damage

8.1 General Introduction

Radiation-induced damage (RID or radiation damage) normally refers to the damage that is due to 'ionizing' damage, typically caused or triggered by neutrons, ions, electrons, and gamma ray sources. The context covers the traditional definition of radiation damage, as defined in ISO 60825, with a wider framework of 'radiation-induced damage'. This book also covers other forms of radiation-induced damage in material processing as well as general radiation failures.

As mentioned in Chapter 2, no FI faced with an unfamiliar case would be able to conduct a satisfactory investigation without knowing the background, which would include the relevant parts of the history of the failed component. Thus, the understanding of radiation sources and damage mechanisms should be well defined and the analysis of such damages should be investigated according to the radiation damage mechanism based on its radiation sources. The new framework in radiation-induced damage is defined as the transfer of energy from an incident projectile (particles or energy) to the solid and the resulting distribution of target atoms after completion of the event. The radiation damage event is composed of several distinctive processes, in which it can be either ionizing or non-ionizing or both.

This chapter will cover the basic theory in these types of radiation damage as well as the key concepts that lead to the damage, in macroscopic and microscopic observation. Three cases have been selected to highlight the theoretical aspects and its failure mechanisms separately. The first, discusses the radiation damage that is caused by typical non-ionizing radiation-induced damage. This is dominantly due to the thermal effect (Sections 8.3 and 8.4). The second, examines ionizing radiation-induced damage that is due to extremely short pulses of energy or a highly concentrated energy packet, that is, ultra-violet (UV) radiation (Section 8.5). The last, as an example, shows higher energy collision, such as high-speed particles, that is, neutrons (Section 8.6).

8.2 Renewed Definition of Radiation-Induced Damage

Laser-induced damage (LID) normally refers to the damage on optical materials caused by laser ablation. LID is generally not perceived under the categories of radiation-induced damage (RID), but rather referred to as the damage that is due to ionizing damage. Typically, RID is triggered by neutrons, ions, electrons, and gamma ray sources. The context covered in this section will not just consider the traditional definition of LID, used in ISO 60825 [1], but also establish a wider framework to define radiation-induced damage that includes some other LID effects, such as ionization mechanisms.

8.2.1 Ionizing Radiation-Induced Damage

Unlike the conventional definition of ionizing radiation, the ionizing energy transferred may have various impacts on the damaging materials, such as lattice dislocation, ionization, and nucleus transformation (transmutation). For example, a femto-second laser can also generate ionization effects, this is called *photo-ionization*; which is exactly the ionizing of damage material using visible/non-visible wavelengths. Generally, laser-induced damaged is considered to be non-ionizing radiation by the International Commission on Non-Ionizing Radiation Protection (ICNRP). This contradicts the conventional understanding of ionizing damage. It is suggested that ionizing RID should be defined as the damage that is caused by higher energy interactions between the radiation sources to the targets. The damaging material shows the ionization effect with its subsequent damage mechanism. This includes high-energy neutrons, deep ultra-violet (UV) radiation, femto-second laser ablation, and so forth. The ionizing effects can be sub-divided into three major forms, namely ionization, change of electronic density states, and transmutations.

8.2.2 Non-Ionizing Radiation-Induced Damage

In this chapter, non-ionizing RID is defined as the damage that is caused by lower energy sources to the targets. The damaging material will exhibit a thermal effect, which includes thermal neutron collision, nano-second laser heating, and so on.

By using the Ullmaier et al. [2] approach, the mechanism can be adopted in this chapter to describe non-ionizing radiation-induced damage conditions. The safety and operational issues related to laser operation is important for this mechanism. Referring to Table 8.1; the mechanism of non-ionizing radiation-induced damage can be defined in the following manner:

1. The interaction of an incident particle/energy packet with a lattice atom;

TABLE 8.1

Non-Ionizing Radiation-Induced Damage Refers to the Damage That Is Caused by Thermal Effects

Time Range (s)	Event
10^{-18}	Energy transfer from the incident particle or energy
10^{-13}	Displacement of lattice atoms
10^{-11}	Energy dissipation, spontaneous recombination, and clustering
10^{-8}	Defect reactions by thermal migration

Source: Approximation from Ullmaier et. al., *Radiation Damage in Metallic Reactor Materials, Physics of Modern Materials*, Vol. I, Vienna: IAEA, 1980 [2].

2. The transfer of kinetic energy to the lattice atom, giving birth to a primary knock-on atom (PKA);

3. The displacement of the atom from its lattice site and dissipation of energy into the lattice.

As indicated in Table 8.1, the non-ionizing RID mechanism dissipates energy at a range larger than 10^{-11} second. This implies that the energy source has to be slower before the dissipation of energy can take place. Specifically, we will focus on laser-induced damage that is above 100 ps pulse length. In order to reach a thermal damage condition, typically with thermal dissipation longer than 10^{-11} second, there is another significant effect that has contributed to the LID mechanism, namely, light intensification.

8.3 Physical Effects of Radiation Damage

As illustrated in Figure 8.1, the proposed framework covers four steps in light intensification and absorption. A typical optical surface can be modelled in the form of crystaline or multi-layer structures. The inherent optical properties, such as the absorption coefficient, can be easily evaluated empirically or from material specifications.

The geometrical condition (morphology) of a given LID site can be updated from its *a priori* distribution or predefined surface topology as an initial condition. For example, Kozlowski and Chow [3] and Genin, Stolz, and Kozlowski [4] modelled the distribution and growth of precursors by increasing the peak fluence of a high power laser and to subsequently quantify how the different types of damage morphologies initiate and grow during repetitive illumination for hafnia-silica multi-layer mirror and polarizer coatings.

Natoli et al. [5] further investigated the agreement between measurement and prediction of different kinds of nano-centres as the precursors of laser

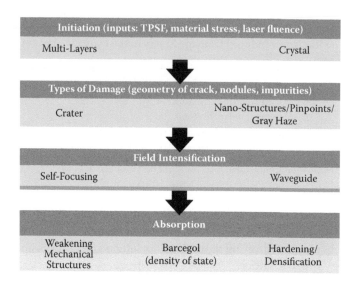

FIGURE 8.1
Non-ionizing LID mechanism.

damage using the probabilistic approach. Hu et al. [6,7] investigated the morphology of 'gray haze' and 'crater'. The test was conducted on the fused silica samples. The gray haze refers to the high-density pinpoint damage morphology. The size of a pinpoint is about 1 μm. The gray haze appears at a fluence higher than ~10 J/cm². The size and depth of the pinpoints that form the gray haze are much smaller than the crater. The difference in the two morphologies is attributed to the property of the absorber and its surrounding material in the redeposition layer, which is different from those in the sub-surface damage layer. Hu et al. [6,7] managed to model the probabilistic distribution of gray haze as a function of the laser fluence rate.

Liao et al. [8] used the Monte Carlo approach to model the evolution of a population of damage sites. The accuracy of the model is compared against laser damage growth observations. In addition, a machine learning (classification) technique independently predicts site evolution from patterns extracted directly from the data. The results show that both the Monte Carlo simulation and machine learning classification algorithm can accurately reproduce the growth of a population of damage sites at a specific fluence range.

The LID initiations start from a precursor [9]. Thus, it is crucial to model the optical material with its initial defects: polishing residues, fractures, contaminations, voids, grain boundaries, and prestressed points (thermal and mechanical). Specifically, for coating materials: the defects could be nodules, substrate contaminants, impurities, cracks. For crystal: atomic impurities, dislocation, and photo-induced defects. The understanding of laser-induced damage in crystal [11–17] is of great importance for technological applications. The damage is initiated by small absorbers that are present in the crystal before irradiation or are formed at the beginning of irradiation. In this context, the damage is created as soon as the inclusion has reached the critical temperature. Recently, Feit and Rubenchik [19] proposed an explanation of overall results based on the analogous work of Hu [20]. They introduced the concept of damage enhancement factor earlier to quantitatively assess the effect of beam modulation (e.g., caused by diffractive optics) on initiation of laser damage. When surface damage is due to an underlying distribution of extrinsic initiators, it can be characterized by the cumulative density, which is the number of initiators per unit area. Typically, the cumulative density is a strong function of laser fluence, for example, a power law or exponential function. The beam hot spots then cause disproportionately more damage than expected from the intensity distribution. However, if the area over which the hot spot occurs is small, it may still be unimportant. This model remains qualitatively valid and is a good starting point for the understanding of laser-induced damage in crystals.

It is assumed that damage is initiated by laser light absorption in nanoparticles. When the particle temperature reaches the critical value Tc, a thermal explosion takes place. The material surrounding the particle becomes absorbing and is transformed into micro/nano growing plasma fireballs. The release of the energy stored in the fireball produces microscopic damage. To determine the laser fluence needed to heat the particle up to the critical temperature, it has to satisfy the heat diffusion equation:

$$\frac{\partial T}{\partial t} = \kappa \nabla^2 T \tag{8.1}$$

where $\kappa = k_m/\rho_m C_m$ (κ, k_m, ρ_m, and C_m are the thermal diffusivity, thermal conductivity, density, and specific heat of the crystals (i.e., KDP). The boundary conditions are (i) at $t = 0$, $T = T_0$ = constant, where T_0 is the initial temperature of the crystal; (ii) at infinity $T = T_0$, and (iii) the power absorbed by the particle must equal the power leaving the surface plus the rate of change of the heat content of the particle,

$$IQ_{abs}\pi a^2 = -4\pi a^2 k_m \left(\frac{\partial T}{\partial r}\right)_{r=a} + \frac{4\pi}{3} a^3 \rho_p C_p \left(\frac{\partial T}{\partial t}\right)_{r=a}, \tag{8.2}$$

where ρ_p and C_p are the density and specific heat of the particle, I is the laser intensity, and Q_{abs} is the absorption efficiency factor and can be calculated from the Mie scattering theory. In Equation (8.2), it is assumed that the temperature of the particle is uniform throughout its volume and equal to the particle temperature to the temperature of the matrix at the surface of the particle. This assumption is justified due to the high thermal diffusivity of the particle.

8.4 Light Intensification due to the Backscattering Effect

Among all the thermal effects due to laser-induced damage, light intensification is considered the most significant factor in the radiation-induced damage mechanism. In order to model the light intensification, it is common to use electromagnetic simulation software to evaluate the electric field intensification. Various attempts have been deployed to assess the relationship between a light intensification with a given defect geometry. Simulation packages, such as using the finite difference time domain (FDTD) method, are used to examine the overall distribution of EM field within the medium and its time domain propagation. Subsequently, the temperature and stress distribution can be evaluated within the medium. However, the actual phenomena from the simulated output are not explained well. Thus, it makes the evaluation of light intensification limited to a particular case of study.

In this section, general backscattering with its implication in light intensification is introduced. The proposed model of reflectance and interference can then be used to explain and predict the light intensification. The material thickness is given by D', where $D' = ml$. l is the unit thickness. Thus, for a layer j (1, 2, 3,....,m), with incoming pulse P_o, the reflectance can be defined as:

$$R_j(t') = \sum_i \int_{t_j}^t S_{ij}\left(t - \frac{2jl}{c}\right) P_0(t' - t)dt \quad for\ t \le \left(\frac{2D'}{c} + t_0 + \langle \tau \rangle\right), \quad (8.3)$$

where t_j is the first mean arriving time of the photon group to layer j. i is the photon history grouping, $i \in$ (DBD, GBD, DBG, GBG, DBM, GBM, MBD, MBG, MBM). Due to its finite thickness D, the convolution between the outgoing pulse P_0 and the kernel function, $S_{ij}(t) = T_{ij}(t)\beta_j(l)$ that is always limited by $t \le \left(\frac{2D'}{c} + t_0 + \langle \tau \rangle\right)$. T_{ij} is the transmittance of the particular photon group, and $\beta(l)$ is the reflectance coefficient as defined by the Fresnel equation. However, if it is a crystalline medium, $\beta(l)$ can be defined as the phase function of the medium at the point of scattering.

For each photon group, the coherency can be correlated based on its mean optical path length. D' refers to the direct photon travelling path, B stands for the backscattered photon direction as defined by its phase function of the medium. A Monte Carlo approach can be used to determine the next backscattered direction based on its cumulative density function of the angular distribution. G is the forward-scattered photon that is also defined by the phase function. Lastly, the M components refer to the multiple-scattered components. In order to simplify the mean optical path length, only the DBD and GBD components are considered in the optical mean free path estimation in Equations (8.3) through (8.6).

The incoming laser pulse, P_j at layer j, can be defined as:

$$P_j(t') = P_0\left(t' - \frac{jl}{c}\right)T_j(t').$$ (8.4)

Subsequently, the light intensification $E_j(t')$ within the coherence length of the laser in layer j is given by:

$$E_j(t') = P_j(t') + R_j(t') + 2\sqrt{P_j(t')R_j(t')}\cos\delta,$$ (8.5)

where σ is the phase difference of incoming and backscattering photons that can be treated separately for each photon group using the mean optical travelling time $<t'>$.

Therefore, it is derived as:

$$OPL_j(t') = \sum_j \left(\frac{c}{n_j}\right)\langle t'\rangle_j \quad \delta = 2\pi\frac{OPL_j}{\lambda}.$$ (8.6)

For the optical medium of LID with finite thickness D', the convolution between the outgoing pulse P_0 and the kernel function, $S_{ij}(t) = T_{ij}(t)\beta_j(l)$ that is always limited by its upper limit, t is less than $(2D'/c + t_0 + <\tau>)$. To further illustrate the intensification criteria, Figure 8.2 only uses the DBD (Direct-Backscattered-Direct) photon group to depict the actual optical path difference over discrete layers of the medium. It has been observed that the light intensification is always stronger at the surface layers compared to the back end of the medium.

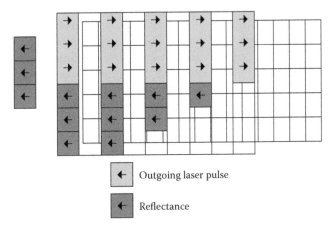

FIGURE 8.2
Light intensification of backscattered and incoming photons over a finite thickness in the optical material (DBD only).

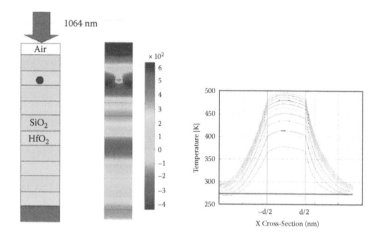

FIGURE 8.3
(See Colour Insert.) Light intensification of backscattered and incoming photons over a finite thickness in Commandre et al. in 2010 [10].

It may be arguable that this is always true at the surface as the laser energy intensification is the highest at the surface. However, the intensification is found to be higher in the sub-layer in the transparent medium and also in the medium that has nodule effects. The nodule is supposed to focus the laser beam towards the backside of the medium. The empirical and simulated results actually observed that the intensification is the highest at the sublayers above the nodule layer. Thus, it is important to evaluate the intensification on the layers. The overall distribution of laser direct heating and the light intensification effect can be evaluated using Equations (8.3) through (8.6).

Some examples are used to illustrate the distribution for this preliminary study of light intensification problems (Figures 8.3 and 8.4). In Figure 8.3, Commandre et al. [10] highlighted the field intensity distribution over a layered medium with a defect at layer 2. The figure indicates a brighter colour for high intensities in Figures 8.3 and 8.4. Layer 5 indicates higher intensities due to its upper limit convolution effect when the pulse passes through the defect. While in layers 7 and 8 it is most likely due to DBM, GBM, MBD, MBG, and MBM components that exhibit higher intensities at lower layers.

In Figure 8.4, the nodule effect is supposed to focus the laser pulse towards the nodule. However, the intensity distribution has concentrated on the surface layer (at 1.3 um). This indicates strong backscattering convolution with an upper limit that interferes with the incoming pulses.

In terms of bulk material temperature estimation, Hu et al. modelled the temperature expression, given as [20,21]:

$$T(r,t) = T_0 + \frac{2_a T_\infty}{r\sqrt{\pi}} \int_\beta^\infty \left\{ 1 - \exp\left[-\frac{t}{\tau_p}\left(1 - \frac{\beta_k^2}{\mu^2} \right) \right] \right\} \cdot \exp\left(-\mu^2\right) d\mu , \qquad (8.7)$$

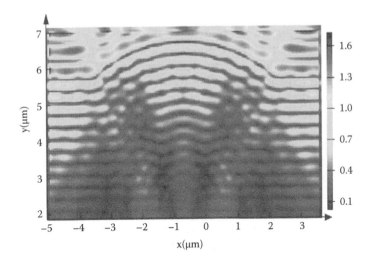

FIGURE 8.4
(See Colour Insert.) Light intensification simulation using Lumerica™.

where $T_\infty = IQ_{abs}a/4k_m$ is the eventual temperature of the particle above the ambient temperature, $\tau_p = a^2\rho_pC_p/3k_m$ is the time to reach it, and $\beta_k = (r - a)(4\kappa)^{1/2}$ is a dimensionless parameter. Equation (8.7) gives the temperature of the matrix at any time t and at any distance r from the centre of the particle due to the heating of this single particle.

The particle temperature as a function of absorber size is plotted in figures published by Hu et al. [40]. This model provides the expressions of pinpoint density and damage probability in agreement with their experimental results.

The temperature rise reached by the most susceptible absorbers heated by a rectangular pulse length, τ_r is given by

$$\Delta T = T - T_0 = \frac{F}{82.5\sqrt{\rho_pC_p\tau_r}}. \tag{8.8}$$

It is assumed that damage takes place when the absorber temperature reaches the critical value T_c. For the (D)KDP characteristic, $\Delta T = T_c - T_0 \sim 4230$ K. It leads to the critical fluence F_c derived from Equation (8.8),

$$F_c = 3.5\times10^5\sqrt{\rho_pC_p\tau_r}. \tag{8.9}$$

For multi-layer coating, the experimental measurements validate a decrease in laser resistance with increased inclusion diameter, as predicted by the normal incident light intensification model [27]. The research also expands the understanding of coating-inclusion-induced light intensification to oblique incidence [28]. Deeply embedded inclusions on the substrate surface were shown to have significant light intensification, consistent with experimental observations that multi-layer coatings are fluence limited by deeply embedded nodules with inclusion diameters exceeding 0.5 μm [5]. Significantly different, light intensification profiles were shown to exist for wavelength differences within the reflection and transmission band [28,30]. Wang et al. [31] extended electric-field modeling of nodular defects into the mid-infrared (3.8 μm). They observed higher light intensification with increased inclusion diameter, decreasing depth, and increased incident angle.

Generally, there are five observations in multi-layer coating LID. The general observations are the following:

- Light intensification increases with inclusion diameter.
- Shallow inclusions tend to have a higher intensification, although deep defects can also have high intensification.
- Incidence angle affects light intensification.
- Light intensification focal spots shift away from the central axis with increasing incidence angle.
- Wavelength changes (either bracketing the reflection band or located within the transmission band) have significantly different electric-field profiles for the same defect geometry and also significantly different light intensification magnitudes.

In addition to the thermal effect, linear and multi-photon absorption also increase with shorter wavelengths, leading to a lower damage threshold. One potential contributing mechanism that has not been discussed here is photon ionization. In the next section, the radiation-induced damage caused by photo-ionization, in high-energy photons, or extra peak power is discussed.

8.5 Photon Ionization due to the Backscattering Effect

This section refers to the work done mainly by Emmert et al. [41], specifically for extremely short pulses of radiation-induced damage, for instance, femto-second pulses. Other forms of photon ionization also include the highly concentrated energy packets, such as in ultra-violet (UV) radiation. The femto-second (fs) laser damage studies on wide bandgap materials indicate that the shorter pulse radiation-induced damage or highly concentrated energy are mainly due to the conduction-band electron density transition. The basic understanding of this form of failure mechanism lies in the band-to-band electron state transition that leads to multi-photon ionization (Keldysh effect), and stronger absorption of energy.

Generally, this is a broad variety of natural phenomena, from corona discharge to thunder lightning, as well as man-made electrical devices such as laser printers and disruptive components in a circuit. Various models have been made available to explain the randomness and fractal nature of a breakdown. The material breakdown model typically describes the macroscopic behavior combining diffusion-limited aggregation within the materials. These phenomenological descriptions fail to reveal the fundamental physical mechanism responsible for an insulating material transforming into a conductive path locally at the breakdown site.

Under strong energy excitation over the electron state, as shown in Figure 8.5, the band-to-band electron transition happens in the multi-photon ionization or impact ionization. Depending on the state of ionization, some of the electrons might leave the surface, where the positively charged lattice results in a Coulomb explosion. Alternatively, the high conduction band (CB) electron densities become unstable, which leads to material bonding failure. The main processes as depicted in Figure 8.5, of photon ionization are due to either multi-photon ionization or impact ionization. Multi-photon ionization is in general an interplay of multi-photon absorption and tunneling ionization as described by the theory of Keldysh [42]. Being multi-photon processes, the electronics in the valence band are elevating their energy states

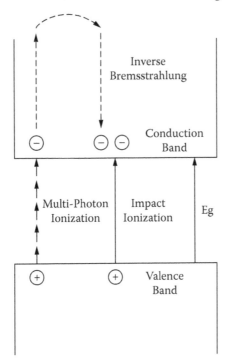

FIGURE 8.5
Photon ionizations due to either multi-photon ionization or impact ionization.

up to the conduction band. While in the conduction band, the electrons can acquire further energy from the laser pulse in a process known as inverse Bremsstrahlung. In the case of fs pulse width, the pulse width is shorter than the vibrational relaxation time constant of several pico-second as mentioned in Section 8.2. As a result, the energy transfer in the conduction band generates free electrons that leave the material surface. The material location forms a plasma state within this process before the electron returns back to the conduction band again. This inverse condition results in a permanent failure of the materials, if the inverse Bremsstrahlung effect has an irreversible effect on the material locally.

8.6 Radiation Damage due to High-Speed Particles

The majority of ionizing radiation damage cases and the literature correspond to the interaction between a high-energy particle and a solid that produces displacement, and consequently, leads to the deformation and failure of the materials. The simplest model is one that approximates the event as colliding hard spheres with displacement occurring when the transferred energy is high enough to knock the struck atom off its lattice site. In addition to energy transfer by hard-sphere collisions, the moving atom loses energy by interactions with electrons, the Coulomb field of nearby atoms, the periodicity of the crystalline lattice, and so on. The problem is reduced to the following key concepts. If one can describe the energy-dependent flux of the incident particle and the energy transfer cross-sections (probabilities) for collisions between atoms, then one can quantify the PKA production in a differential energy range and utilize this to determine the number of displaced atoms. This section will only highlight the elastic case of high-speed neutron–nucleus collisions.

By virtue of their electrical neutrality, elastic collisions between neutrons and nuclei can be represented as colliding hard spheres. When neutrons pass through a solid, there is a finite probability that they will collide with a lattice atom, imparting recoil energy to the struck atom. This probability is defined by the double differential scattering cross-section (in energy and angle), $\sigma_s(E_i, E_f, \Omega)$, where E_i and E_f are the incident and final energies and Ω is the solid angle into which the neutron is scattered. The scattering probability is a function of E_i and the scattering angle. The single differential scattering cross-section is given by Was [43]:

$$\sigma_s(E,\theta) = \int \sigma_s(E_i, E_f, \theta)\, dE_f. \qquad (8.10)$$

The total scattering probability for neutrons of energy E_i is:

$$\sigma_s(E_i) = \int \sigma_s(E_i, \theta) d\theta. \tag{8.11}$$

In the study of irradiation effects, the energy (E) will be transferred from the travelling particle, for instance, from the neutron, to the atom of mass M. This section focuses on elastic scattering where the momentum and energy are conserved before and after the collision. In the binary elastic collisions in the centre of mass frames, the energy transfer to the atom of the mass is defined as:

$$E = \frac{2mME_i}{(M+m)^2}(1-\cos\phi), \tag{8.12}$$

where m = neutron mass, M = target atom mass, and ϕ = the angle of particle movement before and after the collision. From Equation (8.12), it is obvious that the atom of the mass will suffer a dislocation state if $E > E_c$ where E_c is the critical valence bonding strength of the targeted atom.

8.7 Summary

A renewed framework of radiation-induced damage (RID) is introduced in this chapter. Today's technology and knowledge regarding RID, does not make a distinct difference between the mechanism of induced damage. Rather, it is perceived from the point of the radiation source, either in an ionizing or non-ionizing source. In addition, the traditional non-ionizing source, such as a visible light laser, can also induce ionizing damage. This chapter explains the traditional definition of LID, used in ISO 60825. This book proposes a new understanding in general radiation-induced damage that establishes a wider framework to define radiation-induced damage that includes some other LID effects, such as an ionization mechanism.

Unlike the conventional definition of ionizing radiation, ionizing radiation can have various impacts on damage materials, such as lattice dislocation, ionization, and nucleus transformation (transmutation); the radiation damage that is caused by higher energy interactions between the radiation sources to the targets. The damaging material has exhibited the ionization effect with its subsequent damage mechanism. This includes high-energy neutrons, deep ultra-violet (UV) radiation, femto-second laser ablation, and so on. The ionizing effects can be sub-divided into three major forms, namely, ionization, change of electronic density states, and transmutations.

Problem 8.1

From Equation (8.7), notice that the temperature difference over thermal heating by laser pulses can be inversely proportional to the pulse length, τ. Estimate the temperature needed to melt the surface layer of aluminum and state the requirement for pulse length using a 10 J/cm^2 laser?

Answer 8.1

$\rho_p C_p$ of Al is 2.44×10^6 J.m^{-3}K^{-1}. Al melting point is 660°C. Using Equation (8.8), the temperature gradient can be achieved by $F = 10 \times 10000$ J/m^2 is: 1.5 µs, assuming the most susceptible locations on the surface are surface defects that absorb energy using Carr's model.

Problem 8.2

As the application engineer, you are required to examine the effects of UV radiation on the performance and properties of polymer-based building products. Suggest a technique to assess the durability of building materials.

Answer 8.2

Most commercial organic-based polymers used in the building and construction industry undergo certain forms of photo-ionization (photo-oxidation) reactions during exposure to solar UV radiation. The polymers contain chromophoric groups, such as carbon-carbon double bonds (C=C) and carbonyl groups (C=O), which are capable of absorbing UV energy and are involved in the photo reactions that result in the degradation of the polymer.

Some of the plastics, such as polyolefins, polyethylene, and polypropylene, have no carbonyl or C=C bonds within their regular structure, the chromophoric moieties are introduced into the backbone or side groups through the high-temperature injection moulding and extrusion processes.

In order to assess the UV-induced failure, both macro- and micro-examination are useful. The macro-examination will check the cracking on the surface of the materials. Microstructure cracking in prestressed conditions will be good evidence in the assessment of UV-induced failures.

Problem 8.3

For a given particle collision condition that leads to radiation-induced damage, the colliding particle mass, m, to the targeted mass, M, can be related by: $\tan \theta = (M/m) \sin \varphi / [1 + (M/m)\cos \varphi]$, where m and M are the masses of the projectile and target, the scattering angles θ (of colliding particles from the colliding plane in the laboratory space), and ϕ (of colliding particles from the centre-of-mass [CM] frames). Discuss this expression for the following three cases: $m = M$, $m \ll M$, and $m \gg M$.

Answer 8.3

Case $m = M$, $\tan\theta = \sin\phi/[1 + \cos\phi]$. The relation between the CM angle of scattering and the laboratory space-scattering angle is affine linear.

Case $m \ll M$, the scattering becomes elastic. The colliding particles will scatter, which conserves the energy and momentum of the collision with the scattering angle between the laboratory space and the CM system working as a unified system.

Case $m \gg M$, the scattering becomes an inelastic scattering condition; its scattering angle will largely depend on the targeted mass to the colliding mass ratio.

Problem 8.4

In particle physics, a particle with mass, m, is always in a motion state within a predefined dimension, a. This means that the particle is always in a motion state even if it is in a kinetically static condition. The state 1 (minimum) energy level can be described as:

$$E_1 = \frac{\pi^2 \hbar^2}{2ma^2},$$

where $\hbar$ is the plank coefficient.

(a) Calculate the first order of magnitude of the energy levels in atoms and in nuclei using this concept. Use 10^{-10} m for the dimension of the atom and 10^{-15} m for the dimension of the nucleus.

(b) Compared to a free particle, what is the relationship in their energy transfer mechanism?

Answer 8.4

(a) Atomic energy levels: ≈ 40 eV; nuclear energy levels: ≈ 400 MeV.

(b) The free particle will exhibit continuous energy transfer according to the change in its kinetic energy, while a static particle will have a discrete energy transfer depending on its state of energy level and it is inversely proportional to the quantum dimension, a.

References

1. Repocholi, M.H., J.R.A. Lakey, C.B. Meinhold, and C.J. Huyskens, Charter, International Commission on Non-Ionizing Radiation Protection, International Commission on Non-Ionizing Radiation Protection, 1992.
2. Ullmaier, H. and W. Shilling, Radiation Damage in Metallic Reactor Materials, *Physics of Modern Materials*, Vol. I, Vienna: IAEA, 1980.
3. Kozlowski, M.R. and R. Chow, Role of Defects in Laser Damage of Multi-layer Coatings, in *Proc. SPIE of Laser-Induced Damage in Optical Materials*, Ed., H.E. Bennett, L.L. Chase, A.H. Guenther, B.E. Newnam, and M.J. Soileau, *Proc. SPIE*, Vol. 2114, 1994, 640.
4. Genin, F.Y., C.J. Stolz, and M.R. Kozlowski, Growth of Laser-Induced Damage during Repetitive Illumination of HfO2/SiO2 Multilayer Mirror and Polarizer Coatings, in *Proc. SPIE of Laser-Induced Damage in Optical Materials*, Ed., H.E. Bennett, A.H. Guenther, M.R., Kozlowski, B.E., Newnam, and M.J. Soileau, *Proc. SPIE*, Vol. 2966, 1997, 273.
5. Natoli, J.-Y., L. Gallais, H. Akhouayri, and C. Amra, Laser-Induced Damage of Materials in Bulk, Thin-l m, and Liquid Forms, *Applied Optics*, Vol. 41, 2002, 3156–3166.
6. Hu, G., Y. Zhao, D. Li, Q. Xiao, J. Shao, and Z. Fan, Studies of Laser Damage Morphology Reveal Subsurface Feature in Fused Silica, *Surface and Interface Analysis*, Vol. 42, No. 9, 2010, 1465–1468.
7. Hu, G., K. Yi, X. Liu, Y. Zhao, and J. Shao, Growth Mechanism of Laser-Induced Damage in Fused Silica, in *Proc. SPIE of Laser-Induced Damage in Optical Materials*, Ed., G.J. Exarhos, V.E. Gruzdev, J.A. Menapace, D. Ristau, and M.J. Soileau, *Proc. SPIE*, Vol. 8190, 2011, 819020.
8. Liao, Z.M., G.M. Abdulla, R.A. Negres, D.A. Cross, and C.W. Carr, Predictive Modeling Techniques for Nanosecond-Laser Damage Growth in Fused Silica Optics, *Opt. Express*, Vol. 20, No. 14, 2012, 15569–15579.
9. Feit, M.D. and A.M. Rubenchik, Influence of Subsurface Cracks on Laser-Induced Surface Damage, in *Proc. SPIE of Laser-Induced Damage in Optical Materials*, Ed., G.J. Exarhos, A.H. Guenther, N. Kaiser, K.L. Lewis, M.J. Soileau, and C.J. Stolz, *Proc. SPIE*, Vol. 5273, 2004, 264.
10. Commandre, M., G. Demsy, X. Fu, and L. Gallais, Three-Dimensional Multiphysical Model for the Study of Photo-Induced Thermal Effects in Laser Damage Phenomena, in *Proc. SPIE of Laser-Induced Damage in Optical Materials*, Ed., G.J. Exarhos, V.E. Gruzdev, J.A. Menapace, D. Ristau, and M.J. Soileau, *Proc. SPIE*, Vol. 7842, 2010, Section 78420Q.
11. De Yoreo, J., A. Burnham, and P. Whitman, *Int. Mater. Rev.*, Vol. 47, 2002, 113.
12. Burnham, A., L. Hackel, P. Wegner, T. Parham, L. Hrubesh, B. Penetrante, P. Whitman, S. Demos, J. Menapace, M. Runkel, M. Fluss, M. Feit, M. Key, and T. Biesiada, *Proc. SPIE*, Vol. 4679, 2002, 173.
13. Feit, M., A. Rubenchik, and M. Runkel, *Proc. SPIE*, Vol. 4347, 2001, 383.
14. Carr, C., M. Feit, and A. Rubenchik, *Proc. SPIE*, Vol. 5991, 2005, 59911Q.
15. Hu, G.H., Y.A. Zhao, S.T. Sun, D.W. Li, X.F. Liu, X. Sun, J.D. Shao, and Z.X. Fan, *Chin. Phys. Lett.*, Vol. 26, 2009, 097802.

16. Hu, G.H., Y.A. Zhao, S.T. Sun, D.W. Li, X.F. Liu, X. Sun, J.D. Shao, and Z.X. Fan, *Chin. Phys. Lett.*, Vol. 26, 2009, 087805.

17. Hu, G.H., Y.A. Zhao, S.T. Sun, D.W. Li, X. Sun, J.D. Shao, and Z.X. Fan, *Chin. Phys. Lett.*, Vol. 26, 2009, 087801.

18. Hopper, R. and D. Uhlmann, *J. Appl. Phys.*, Vol. 41, 1970, 4023.

19. Feit, M. and A. Rubenchik, *Proc. SPIE*, Vol. 5273, 2004, 74.

20. Hu, G.H. et al., Characteristics of 355 nm Laser Damage in KDP and DKDP Crystals, *Chinese Phys. Lett.*, Vol. 26 097802, 2009.

21. Carslaw, H.S. and J.C. Jaeger, *Conduction of Heat in Solid*, 2nd ed., Oxford: Oxford University, 1959.

22. Carr, C., H. Radousky, A. Rubenchik, M. Feit, and S. Demos, *Phys. Rev. Lett.*, Vol. 92, 2004, 87401.

23. Dyan, A., M. Pommies, G. Duchateau, F. Enguehard, S. Lallich, B. Bertussi, D. Damiani, H. Piombini, and H. Mathis, *Proc. SPIE*, 2007, 640307.

24. Feit, M., A. Rubenchik, and J. Trenholme, *Proc. SPIE*, Vol. 5991, 2005, 59910W.

25. DeFord, J.F. and M.R. Kozlowski, Modeling of Electric-Field Enhancement at Nodular Defects in Dielectric Mirror Coatings, in *Laser-Induced Damage in Optical Materials*, Ed., H.E. Bennett, L.L. Chase, A.H. Guenther, B.E. Newnam, and M.J. Soileau, *Proc. SPIE*, Vol. 1848, 1993, 455–470.

26. Sawicki, R.H., C.C. Shang, and T.L. Swatloski, Failure Characterization of Nodular Defects in Multi-Layer Dielectric Coatings, in *Laser-Induced Damage in Optical Materials*, Ed., H.E. Bennett, A.H. Guenther, M R. Kozlowski, B.E. Newnam, and M.J. Soileau, *Proc. SPIE*, Vol. 2428, 1995, 333–342.

27. Kozlowski, M.R., R.J. Tench, R. Chow, and L. Sheehan, Influence of Defect Shape on Laser-Induced Damage in Multiplayer Coatings, in *Optical Interference Coatings*, Ed., F. Abelès, *Proc. SPIE*, Vol. 2253, 1994, 743–750.

28. Stolz, C.J., M.D. Feit, and T.V. Pistor, Laser Intensification by Spherical Inclusions Embedded within Multilayer Coatings, *Appl. Opt.*, Vol. 45, 2006, 1594–1601.

29. Dijon, J., M. Poulingue, and J. Hue, Thermomechanical Model of Mirror Laser Damage at 1.06 μm. Pt. 1: Nodule Ejection, in *Laser-Induced Damage in Optical Materials*, Ed., G.J. Exarhos, A.H. Guenther, M.R. Kozlowski, K.L. Lewis, and M. J. Soileau, *Proc. SPIE*, Vol. 3578, 1999, 387–396.

30. Stolz, C.J., F.Y. Génin, and T.V. Pistor, Electric Field Enhancement by Nodular Defects in Multilayer Coatings Irradiated at Normal and 45° Incidence, in *Laser-Induced Damage in Optical Materials*, Ed., G.J. Exarhos, A.H. Guenther, N. Kaiser, K.L. Lewis, M.J. Soileau, and C.J. Stolz, *Proc. SPIE*, Vol. 5273, 2004, 41–49.

31. Wang, Y., Y. Zhang, X. Liu, W. Chen, and P. Gu, Gaussian Profile Laser Intensification by Nodular Defects in Mid-Infrared High Reflectance Coatings, *Opt. Commun.*, Vol. 278, 2007, 317–320.

32. Bodemann, A., N. Kaiser, M.R. Kozlowski, E. Pierce, and C.J. Stolz, Comparison between 355 nm and 1064 nm Damage of High-Grade Dielectric Mirror Coatings, in *Laser-Induced Damage in Optical Materials*, Ed., H.E. Bennett, A.H. Guenther, M.R. Kozlowski, B.E. Newnam, and M.J. Soileau, *Proc. SPIE*, Vol. 2714, 1996, 395–404.

33. Pistor, T., Electromagnetic Simulation and Modeling with Applications in Lithography, Ph.D. dissertation, Berkeley: University of California at Berkeley, 2001.

34. Lowdermilk, W.H., D. Milam, and F. Rainer, Damage to Coatings and Surfaces by 1.06 µm Pulses, in *Laser-Induced Damage in Optical Materials*, Ed., H.E. Bennett, A.J. Glass, A.H. Guenther, and B.E. Newnam, *NIST Spec. Publ.*, Vol. 568, 1980, 391–403.

35. Tench, R.J., R. Chow, and M.R. Kozlowski, Characterization of Defect Geometries in Multilayer Optical Coatings, *J. Vac. Sci. Technol.* A, Vol. 12, 1994, 2808–2813.

36. Stolz, C.J., R.J. Tench, M.R. Kozlowski, and A. Fornier, A Comparison of Nodular Defect Seed Geometries from Different Deposition Techniques, in *Laser-Induced Damage in Optical Materials,* Ed., H.E. Bennett, A.H. Guenther, M.R. Kozlowski, B.E. Newnam, and M.J. Soileau, *Proc. SPIE*, Vol. 2714, 1996, 374–382.

37. Borden, M.R., J.A. Folta, C.J. Stolz, J.R. Taylor, J.E. Wolfe, A.J. Griffin, and M.D. Thomas, Improved Method for Laser Damage Testing Coated Optics, in *Laser-Induced Damage in Optical Materials,* Ed., G. Exarhos, A.H. Guenther, K.L. Lewis, D. Ristau, M.J. Soileau, and C.J. Stolz, *Proc. SPIE,* Vol. 5991, 2006, 59912A-1–59912A-8.

38. Göppert-Mayer, M., Uber Elementarakte mit zwei Quantensprungen, *Ann. Phys.,* Vol. 9, 1931, 273–295.

39. Walker, T.W., A.H. Guernther, and P. Nielsen, Pulsed Laser-Induced Damage to Thin-Film Optical Coatings—Pt. I: Experimental & Pt. II: Theory, *IEEE J. Quantum Electron,* Vol. QE-17, 1981, 2041–2065.

40. Hu, G.H., Y.A. Zhao, S.T. Sun, D.W. Li, X.F. Liu, X. Sun, J.D. Shao, and Z.X. Fan, A Thermal Approach to Model Laser Damage in KDP and DKDP Crystal, *Chinese Phys. Lett.,* Vol. 26, No. 9, 2009, 097803.

41. Emmert, L.A., D.N. Nguyen, M. Mero, W. Rudolph, D. Ristau, K. Starke, M. Jupe, C.S. Menoni, D. Patel, and E. Krous, Fundamental Processes Controlling the Single and Multiple Femtosecond Pulse Damage Behavior of Dielectric Oxide Films, in *Laser Induced Damage in Optical Materials, SPIE Proc.,* Vol. 7504, 2009, 75040P-1-10.

42. Keldysh, L.V., Ionization in the Field of a Strong Electromagnetic Wave, *Sov. Phys. JETP,* Vol. 20, 1965, 1307–1314.

43. Was, Gary S., *Fundamentals of Radiation Material Science*, New York: Springer Berlin Heidelberg, 2007.

Index